George Wall

The Natural History of Thought in its Practical Aspect

From its Origin in Infancy

George Wall

The Natural History of Thought in its Practical Aspect
From its Origin in Infancy

ISBN/EAN: 9783337026905

Printed in Europe, USA, Canada, Australia, Japan

Cover: Foto ©berggeist007 / pixelio.de

More available books at **www.hansebooks.com**

THE
NATURAL HISTORY OF THOUGHT

BY THE SAME AUTHOR.

GOOD AND EVIL IN THEIR RELATION TO THE DISPENSATIONS OF PROVIDENCE.

London: F. NORGATE. 1886.

"This is a most valuable and interesting little work. .... It is a work to be read, re-read, and pondered over by all our religious instructors. The author possesses the indispensable qualification for guiding us to a right decision on this most important subject; for he seems to be as well versed in God's book of Nature as in that of Revelation. His general views of the Universe in relation to Good and Evil are admirable, and will well repay the most careful study."—*The Church of England Pulpit and Ecclesiastical Review.*

"The author shows that evil, as the correlative of good, fulfils a necessary function in the economy of God's government, and is an essential factor in the Universe. The argument is marked by thoughtfulness and wisdom."—*The Christian World.*

"A small book on a great subject, and yet it contains more vigorous helpful thinking than can be found in many a large and pretentious volume; and the clearness and beauty of the style add greatly to the charm and value of the thought. .... It is a long time since we have met with a book of such fresh, enlarged and helpful thought; it is a noble contribution to Christian faith."—*The Christian Age.*

"This is a most able work. The author has a firm grasp of his subject, and he knows how to expound, defend and illustrate it. The volume is really an invaluable contribution towards the solution of the greatest moral problems of all times."—*Oldham Chronicle.*

THE
NATURAL HISTORY OF THOUGHT

IN ITS PRACTICAL ASPECT

FROM ITS ORIGIN IN INFANCY

BY

GEORGE WALL, F.L.S., F.R.A.S.

AUTHOR OF
"GOOD AND EVIL IN THEIR RELATION TO THE DISPENSATIONS OF PROVIDENCE"

"Keep thy heart with all diligence, for out of it are the issues of life."
Prov. iv. 23

LONDON
TRÜBNER & CO., LUDGATE HILL
1887

[All rights reserved]

Ballantyne Press
BALLANTYNE, HANSON AND CO.
EDINBURGH AND LONDON

PREFACE.

THOUGHT may be regarded in three aspects: as a mysterious element of Nature, as a marvellous interpretation of Nature, or as a treasury of experience of Nature.

In the first-mentioned of these aspects Thought shares in the insoluble mystery of other vital phenomena and eludes research. The second or philosophic aspect has engaged the powers of the deepest thinkers for ages, and has given rise to various theories of most seductive interest, which however have little bearing upon the cultivation of the intellect. The third, or practical aspect, concerns the use and culture of the thinking faculty for the purposes of life.

The philosophic inquiry is suggested by the fact that sensations are not copies but interpretations of Nature; that, in fact, there exists between the inner self and the outer objects of Nature an interpreting faculty which presents those objects to the percipient in a garb not proper to themselves, but to its office as interpreter. Light, for example, which consists of mere vibrations, is interpreted by the eye and presented to the mind in the familiar sensation of luminosity. All the glories of the landscape as they reach the eye are mere vibrations and are not inherent in the objects themselves, but owe their existence to the office of the eye as interpreter of those vibrations. In like manner the other senses in the fulfilment of their several offices effect similar interpretations of external phenomena. Between the clicks of a telegraph and the meaning of a message there is an interpreting telegraphist, but there is no actual resemblance between the clicks and the message. So in like manner the presentations of Nature to the mind through the senses are quite different from the objects of Nature, and hence the question arises as to the actual relation that exists between the percipient and the outer world, and as to the substantial nature of those objects themselves as distinguished from their interpretations. The inquiry of the philosopher is of the highest conceivable interest, but has little, if any, practical bearing upon the culture of the intellect. Hence the philosopher pursues it with intense ardour; but the parent, intent upon the education of his child, finds in it no help for the object he has in view. To him the practical aspect of Thought and its application to the business of life are of far greater moment than its scientific subtleties, and it is to him that the following pages are addressed.

In the practical treatment of this most important and interesting subject in this work, it has seemed necessary sometimes to employ the terminology and to refer to the processes of mental science; but this is done only in so far as they serve the practical purpose in view and without any pretensions to scientific exactness or intention of assuming the functions of the philosopher.

In every other branch of knowledge it is deemed necessary to begin at the beginning, but in regard to the cultivation of the mental faculties this rule is not observed. Nearly all that is done or attempted in the way of training the intellect is applicable only to the mind in the advanced stages of its history. The rules prescribed and the means used in the education of youth are designed to furnish information to minds already more or less developed, and are inapplicable to the incipient faculty. They therefore afford no help towards moulding the mind and directing its first efforts during the plastic period of its formation. If the mind had no incipient stage, but sprang into being at once with a mature power of understanding, there would be reason in regarding it in that aspect, and in providing only for its requirements in that condition. Seeing, however, that it resembles all other of God's works in attaining its powers by a gradual process of development from zero, it seems proper that it should be studied and cultivated from its origin in babyhood. The birth, progress and early development of the faculty being generally disregarded, the infant mind is left to form or deform itself under such influences as may happen to environ the first stages of childhood and supply the first material for the exercise of the incipient intellect.

In the absence of regular training, however, the mind does not wait, but exerts its powers, acquires its force, and receives its bent from such conditions and influences, bad or good, as the surroundings may happen to supply. On these, generally casual, often adverse, and rarely ever designed with any reference to what is or may be going on in the incipient mind, depends the actual and practical education imparted. These conditions and influences determine the nature of the characters first inscribed on the susceptible tablet of the infant mind, and lay the foundation of the superstructure to be built thereupon.

Few persons would be inconsiderate enough to declare in so many words that it matters not what a child of a year old thinks, yet people generally act as if such were their belief. Scarcely any one thinks it worth while to ascertain or even to consider what goes on in the mind of an infant, though many are ready to admit that, in some sense, "the child is father to the man." We are careful, therefore, in this essay to demonstrate how much is really and indisputably accomplished in the way of mental work by every sane child during his first years of life, and by this means to prove the need for assiduous care and attention during that critical and trying period of individual history.

Our object is to trace the birth and progress of the thinking faculty, and to learn the manner of its growth from its earliest dawn to the maturity of its powers, in order to ascertain the proper means by which it may be moulded and directed during its plastic stage. The physical powers require and receive for their development food and exercise so adjusted as to ensure all attainable vigour, and in like manner the mental powers require similar treatment, and should not be allowed to run wild under casual influences, but be subjected to regular culture based upon the principles which determine their development.

Temper, disposition, respect for authority, a power of giving fixed attention, and other essential qualifications for effective learning may be acquired, but they do not come of themselves. If left to chance, as they usually are, these qualities and habits of mind will be as variable as the chances which decide them. Such qualities need to be cultivated in early childhood, from the beginning in fact, by appropriate means suggested by and suitable to the natural conditions of growth. When the systematic education of school life commences, certain conditions of mind and habits of Thought will, in all cases, have been already fixed in one way or another, for good or for evil. If these be adverse, which according to the doctrine of chances they often are, they constitute serious disqualifications for the all-important work of that period. In such case a youth may seem, even when doing his best, to be stupid or perverse. A rap on the knuckles is then expected to subject his Thoughts to instant discipline, to enlighten his understanding, and to do duty for previously neglected training. The work assigned to a pupil cannot be adapted to the instrument by which it is to be performed if the mind has not received suitable preparatory training, but has acquired its form and force under casual influences in infancy and childhood. If the natural laws which determine the nature of the product be ignored and neglected the result will be as uncertain as the causes which produced it. Hence the strange diversity of result which so often confounds the hopes and expectations of anxious parents.

The writing of this work has occupied an interval of forced leisure due to partial and sometimes total blindness. Hence a considerable part of it was written in the dark with mechanical aid, and the whole has been composed without the advantage of reading or even of referring to authorities. We are therefore conscious of much imperfection in the performance, but we have made the best we could of our opportunity, which is now at an end. We must therefore appear under existing disadvantages or not at all. The alternative has been decided by friends, who think the book may serve the useful purpose contemplated by the writer.

CONTENTS.

CHAPTER I.
INTRODUCTORY.

Mysterious nature of Thought, 1—Its practical use independent of metaphysical views, 2—Embodies experience of life and its opportunities, 3—Is the most enduring wealth, 4—Determines what a man is; Thought a constant companion, 5—An index of character, 7—All that remains after death, 8—Makes or mars the happiness of life, 9—Importance of its history, 10—Why is mental training deferred, 11—Kindergarten system; Temper an early acquirement, 12—The work of life to choose between good and evil, 14—Current nursery practice, 15—General plan of the sequel, 16—Thought the result of a process; factors considered separately, 17—Definition of the term "Tutor," 18.

CHAPTER II.
ON THE NATURE OF THOUGHT.

Thought consists of ideas of what is noticed; Attention necessary to formulation of ideas, 19—Factors involved in the production of an idea, 21—Sources of the subject-matter, sensation and reflection, 22—First sensuous and involuntary, 22—Sensations not all formulated, 24—Ideas classified, 25—Simple ideas, 26—Complex ideas, 27—Indeterminate ideas, 28—Illustrated, 29—Influential in babyhood; No innate ideas, 30—Thinking faculty in the lower animals, 31—Affords instructive lessons, 32—Two domains of Thought, 33—Mental and moral, 34—Spiritual life; Buddhistic doctrine, 35—Qualifications for spiritual life, 36—Order of animated beings, 37—Diagram of the order, 38—Conclusion, 39.

CHAPTER III.
ON THE FUNCTIONS OF THOUGHT.

Primary functions for the business of life—ultimate function for reproduction of Thought, 40—Functions in relation to individuals and to sentient beings generally, 41—Order of their operation, the same in both, 42—In the lower animals, 43—Infant begins life in the lower stage, 44—Conditions of babyhood, 45—And first ideas of good and evil, 46—Stages of child-life; their characteristics, 48—And progress of Thought then, 49—And afterwards, 50—The lower domain existed for ages before the higher; purposes of each, 51—Man by his higher nature renders to God voluntary homage and love, 52—And subdues the lower nature, 53—A future state implied in the conditions of the present life, 55—All experience perpetuated in Thought—of the race as of the individual, 56—Ultimate functions of Thought, 57.

CHAPTER IV.

ON LANGUAGE.

Lower animals have rudiments only, 59—Very slowly developed into language by man, 60—Learnt by imitation and inference, 60—Factors involved in understanding speech, 62—And difficulties, 63—Apprehension of the uttered sound, 64—Subject-matter; Association of word and idea; Recollection of both, 65—All these factors necessary, 66—Process becomes mechanical in adults; Further requisites for expressing, 68—Articulation, &c., 68—Written language, 69—Spelling, 70—Alphabet, 71—Uses of language, 72—Men do not always think in words, 73—Use of language in mental work, 74—Its imperfections—Our history divided into periods by progress of language, 75.

CHAPTER V.

ON TEMPER.

Effect of temper on character, 77—And on habit of thinking, 78—Temper originates in feelings, 79—And corresponds thereto, 80—Controlled by adults and partly by children; naturally exhibited by lower animals, 81—Illustrative cases, 83, 84, 85—Suasive and coercive treatment compared, 85—Brutes kind to their young, 86—Natural effects of treatment of infants, 87—Opportunity afforded by conditions of baby life, 88—When feelings predominate, 89—Trials of the toilet, 90—All infant experience influential; and serves important purpose, 91—Crosses abound through life, and most trying in childhood, 92—Temper in its bearing on school work, 93.

CHAPTER VI.

ON THE NATURAL HISTORY OF THOUGHT. FIRST PERIOD: BABYHOOD.

Feelings the dominant influence of the period, 95—Corporeal faculties, 96—Brought under control; the Will, 97—In the brute races, and in the babe, 98—Mental faculties; Personal feelings, how affected, 99—External objects, 100—Evidence of the eye and ear; Music, 101—Ideas formulated in babyhood; Memory, 102—Reflection, 103—Recognition and association, 104—Inference weak; Temper initiated, 105—Faith in tutors the basis of faith in God, 106—Nature of true obedience; Moral obligation, 107—Suggested by example, 108—Habit of Thought; Summary of progress, 109.

CHAPTER VII.

SECOND PERIOD: INFANCY.

Activity the dominant influence, 112—Corporeal faculties, 113—Should be employed, 114—Imitation, articulation, 115—Nervous system, 116—The Will to be moulded; effects of neglect, 117—Mental faculties; Personal sensibilities tried, 118—Effects of treatment, 119—External objects increasingly diverting, 120—Acute discernment, 121—Ideas of injustice, 122—Memory; Comparison and association, 123—In acquiring language, 124—Inference, &c., 125—Locke on infants, 126—Temper needs care, 127—Faith, the basis of true obedience, 128—Fifth commandment, 129—Adam's first lesson; Moral principles, 130—First ideas lasting; truth, 131—Nursery fictions; Recitations, 132—Habit of Thought, especially attention, 133—Summary of progress, 134.

CHAPTER VIII.

THIRD PERIOD: CHILDHOOD.

Inquisitiveness the dominant influence, 137—Corporeal faculties to be employed, 139—The Will now formed, 140—Mental faculties; Learning by language, 141—Reading, 142—Personal feelings, 143—Respect for authority, 144—External objects better understood; criticisms, 145—Memory required for learning language, 146—Number of words understood and used, 147—Conditions favourable for learning, 149—Comparison and association; written language, 150—Thought requires time; personal equation, 151—Inference and judgment the highest faculties, 152—Temper now decided, 153—Effect of teasing; Faith now confirmed or impaired, 154—Importance of consistency; Obedience now necessary; often coerced, 155—Lessons trying, 156—Moral principles by direct tuition, 157—Worship, sermons, 158—Habit, how formed, 159—Effect on mental work, 160—Summary of progress, 161.

CHAPTER IX.

FOURTH PERIOD: YOUTH.

Aspirations for independence the dominant influence, 163—Corporeal faculties; athletics, 164—Nervous system, 165—The Will, effect of discipline, 166—Training of lower animals; natural conditions of each period suited to its work, 167—Willingness necessary to learning, 168—Mental faculties employed on two kinds of subject-matter, personal and external; learning involves both, 169—Educational appliances, 170—Personal feelings change from stage to stage, 171—Sympathy inspires confidence, 172—External objects more attentively regarded, 173—Concentration; interest necessary for school work, 174—Choice of a calling, 175—Memory; nature of the stock of ideas, 176—In relation to language, 177—Course of mental progress, 178—Learning and discipline, 179—Judgment freely and confidently used, 180—Temper now a great help or hindrance, 181—Faith; its forms and effects, 182—Obedience, real or artificial; moral principles practically tested, 184—Habit an exponent of character, 185—Summary of progress, 186.

CHAPTER X.

REVIEW OF THE FOUR PERIODS.

Relation of mental to material history, 190—Of mind to organism, 191—Moral element, 192—Order of progress in man and in nature, 193—Man distinguished from brute by knowledge of good and evil, 194—The higher and lower natures in man, 195—Stages of mental progress; Language, 196—Special conditions affecting mental work—Sex, 198—Temperament, 200—Social grade, 201—Environment and example, 203—Natural affection; family institution, 206—Ignorant nurses, 207—Facial expression, 208—Perfection in tutor's example impossible, 209—And unnecessary; Sickness, infirmity, &c., 210—Teething. Effect of encouragement, 211—And of ridicule, 212.

CHAPTER XI.

ON THE HABIT OF THOUGHT.

Habit insensibly contracted and formed early, 214—Especially of temper, 215—Imitation—Obscure origin of habit, 217—Insensible progress and great influence, 218—Order of formation, 218—Effect of temperament, 219—Habits of listlessness and earnestness, 220—Their effects, 221—In reference to temptation, 222—And to mental power, 223—Victims of listless habit often regard it as inherited, 224—This theory demoralizing, 225—Control to be cultivated, 226—Attention as a habit, 227—Encouraged in infancy, 228—Kindergarten methods effective, 229—Habits suitable for school-work, 230—Of observation, 231—Supplementation of indistinct ideas, 233—Habit in relation to judgment, 234—To teacher and pupil, 235—Absence of mind, 236—Bearing of habit on moral principles, 237—Its relation to different occupations, 238.

CHAPTER XII.

ON THE CONTROL OF THE THINKING FACULTY.

Control subject to the Will, modified by habit, 240—Difficult to attain, 241—Freedom of the Will, 242—A faculty distinct from the organism, 243—First stimulated by natural impulse to use the faculties, 244—Originally under control of tutors, 246—Hence their responsibility, 247—Influences controlling the Will, parental, 248—Neglect of, 249—Coercion, its effect on lower animals, 250—In family discipline, 251—Indian Mutiny. Consideration of consequences; of human laws; of moral principles, 252—Wisdom of moral law as shown in its operation, 253—Education of the Will, 254—In babyhood, 255—And afterwards, 257—Inimical influences; listlessness; indulgence, 258—When unrestrained, 259—Harbouring thoughts of indulgence, 260—Indolence, 261—Dependence on others, 262—As shown in hypnotism, &c., 263—Control in relation to character, 264—Power of choosing currents of Thought, 265—Attainable by every one, 266—Painful reflections often salutary, 267—When not remorseful, 268—Mental pleasure grounds for leisure thought, 269—Control of Thought easy when it is habitual, 270—Practical conclusion, 271.

CHAPTER XIII.

ON MEMORY.

Neither past nor future without memory, 272—Memory depends on the kind of ideas in the mind, 273—And their formulation, 274—Illustrated, 276—Conditions favourable to recollection, 277—Experiences of babyhood, 278—And their effects, 279—Of childhood, 280—Memory recalls ideas as they were formulated, including errors and defects, 281—The index of the mind, 282—Illustrations, 283—Affected by habit of Thought and observation, 284—Conditions which determine the strength of ideas, 285—Attention, 286—Nature and duration of opportunity, 289—Nature of subject-matter, 291—Mental condition at the time of formulating, 292—Pre-occupation; attendant circumstances, 293—Coincidences; an ancient practice, 294—Repetition, 295—Of different kinds, 296—Most effective aid, 297—Mnemonic systems, 298—Their rules, 299—Not always applicable, 300—Ideas to be remembered must be kept in use, 301—Obtrusive ideas—Early training, 302.

CHAPTER XIV.

ON JUDGMENT.

Factors involved in process of judging, 304—(1) Power of the faculty, 305—Accords with strength and sufficiency of evidence, 306—And with habit of thinking, 307—(2) Stock of evidential knowledge—Its extent, value and nature, 308—Extent varies according to age, opportunity, environment and habit of each person and the power of the senses, 308—Effect of age, 309—Opportunity, 310—Environment, 312—Habits of Thought, 313—Efficiency of senses, 314—Value of the stock of evidential knowledge depends on its readiness of recall and its truth, 315—Its readiness, 316—Its truth, 317—Effects of error, bigotry, 318—Superstition, 319—Baconian research, 320—Nature of the fund of evidential knowledge, 321—Particular ideas and general principles and standards, 321—Particular ideas, 322—General principles, 323—Their importance, 324—Standards, their influence, 325—Vary from time to time, 326—(3) Application of evidence to purpose of judging, 327—Principles involved, 328—Need to be taught in youth, 329—Nature and effect of prejudice, 330—The mind should be both disciplined and informed, 331—Euclid, 332—Babbage's illustration, 333—Instance of prejudice, 335—Imperfections of language as a source of error, 336.

CHAPTER XV.

ON INHERITED CAPABILITIES.

Diversity of powers, 339—Ideas not heritable being produced by sensations, 339—No feeling possible anterior to life itself, 340—Reasons for the theory of inherited ideas, 341—Evidence of phrenology, 342—Our five propositions, 343—(1) Life itself not inherited, 343—Essentially an active force, 344—Seeds germinating after four thousand years, 345—Receptivity not less wonderful than life itself, 345—Life in plants, 345—In animals, 346—What is the actuating principle in man, 347—The use, not the possession, of powers determines action, 348—Brain instrumental only, not actuating, 349—Effect of trivial events, 349—In deflecting actuating forces, 349—Illustration, 350—(2) Men do not always use the powers they possess, 352—Sometimes unconscious of their powers, 353—Talent often comes by application, 354—(3) Extraordinary mental attainments, 355—Mozart, 356—Shakespeare, 357—No special powers implied in their attainments, 358—The will actuates, 359—Instances of great attainments by men of ordinary power, 360—Ideas of inherited habits, 360—(4) Inherited qualities as actuating principles opposed to experience, 363—Resemblances in families, 364—And far greater dissimilarities, 364—(5) Natural faculties neither superseded nor anticipated, 365—All begin at zero, 365—Baby conditions suited to initiation of mental work, 366—Functions of the brain, 367—Divine command implies man's power to obey, 368—Evidence of phrenology, 369—Practical effect of the theory of inherited gifts, 370—Illustrative cases, 371.

CHAPTER XVI.

ON THE EARLY TRAINING OF THE MIND.

In babyhood; our method exacts no effort of the babe, 373—Our aim a right beginning; Principles of our method, 374—Conditions recapitulated, 375—Rules, 376—Inconsistency of nursery practice illustrated, 377—Resentment—Useful purpose of the exigencies of babyhood, 378—Thought accords with feelings, 379—The will; Baby always imbibing ideas from example—Effects of shocks and terror, 380—Training in infancy—Joy natural to young creatures, 381—Coercion deprecated, 382—Correction as a principle; Rewards and punishments, 383—God's dealing with Adam; Good and evil produce their natural results, 384—Punishment only deterrent, not corrective, 385—Rule 6, its universal applicability, 386—Imitative propensity illustrated, 387—Infant ideas of justice; Occupation, 388—Training in childhood; one lesson at a time, 389—Lessons and teaching, 390—Most learning is by observation, 391—Attention difficult to keep, 392—Attention cannot be taught, 393—Rule 7, taking advantage of opportunities, 394—Conclusion; Effects of a right beginning, 396—And pursued onwards, 397—Our rules simple and scriptural, 398—Perfection the aim, though not attainable until God's will be done on earth as it is in heaven, 400.

THE NATURAL HISTORY OF THOUGHT

IN ITS PRACTICAL ASPECT.

CHAPTER I.

INTRODUCTORY.

Of all mysteries none, except the supreme mystery, God himself, is greater, more inscrutable, or at the same time more real and influential than the faculty of thinking. Like the sister mysteries of life and Will, its essential nature and mode of operation surpass the utmost powers of human comprehension. All the resources of science and known means of research fail to resolve the impenetrable secret of the intrinsic nature of Thought, or of its mode of action. Nevertheless, like those other mysterious agencies, it may be used and enjoyed in the fullest measure, and be made to realize all the blessings and advantages resulting from its exercise, without any better knowledge, on the part of its possessor, as to its essential nature, than that which every one may learn from his daily experience. The enjoyment of life suffers no abatement whatever from ignorance of the inscrutable nature of its phenomena, and is probably most fully appreciated by those who think least of the mystery that surrounds it. The same may be said of Thought.

We do not propose, therefore, to discuss the metaphysical aspects of Thought, or of the process by which the thinking faculty produces and elaborates it. Deeply interesting as such a discussion might be, it would have no bearing on the purpose we have in view. Even if any elucidation of the mystery were within reach, it would not further the object we contemplate. If it were conclusively demonstrated that certain molecular changes in the tissues of the brain were the effective causes of the phenomena of Thought, neither the resulting product, nor the essential nature of the process of thinking, would be one whit the less inscrutable. If, for example, it were proved that, by some process analogous to photography, certain definite characters were inscribed on the brain by sensations or reflections, these inscriptions would still need to be read and interpreted. Whatever their nature might be, the characters inscribed could not interpret themselves, but would need to be apprehended by some faculty gifted with the power of reading them. The inscrutable goal of the research would still be as far as ever from the grasp of the inquirer. We make no attempt therefore to lift the veil which conceals from human ken the mysteries involved in the production of Thought, or in the operation of the Will in actuating and directing the process of thinking. We confine our discussion to the practical and vitally important sense in which these faculties are universally employed and understood, and in which sense alone they concern the conduct of life.

The faculty of thinking is the highest with which

man is endowed. By its agency he is enabled to formulate into concrete ideas, and fix in the mind, all such of the otherwise transient and fugitive sensations, reflections and experiences of life, as he chooses to gather into the stock of his knowledge. So much, and *only* so much, as is thus formulated and fixed, of all that passes within the sphere of observation, is rescued from oblivion, added to the mental treasury, and incorporated into the mind, as a part of the individual self.

Thought, therefore, is the formulated embodiment or record of all those sensations, reflections, and events of life which are appropriated, and added, more or less permanently, to the stock of experience. It is what remains in possession of all that comes within the range of perception after the observed facts or sensations have themselves passed away. It represents the use that is made of all the opportunities life affords for the accumulation of man's richest treasure. It measures the extent to which the highest talent entrusted to his stewardship is improved; embodies the opinions, principles and judgments that have been elaborated out of all that has come within the range of his consciousness in the past for his use and guidance in the future, and forms the capital by means of which the whole economy of life is conducted. It is the source whence outward actions flow, the spring of the noblest aspirations and also of the basest conceptions, of the most generous impulses as well as of the most sordid designs, of all the good men do or desire to do, and of all the evil they accomplish, as well as of that which fear of consequences or lack of opportunity prevents.

The thinking faculty is naturally subject to the

control of the Will, and always acts under its impulses, except in so far as these may have become crystallized into the form of habit. If, however, Thought be not kept regularly under control, the power of the Will is weakened by disuse, and the mind becomes subject to wandering, desultory Thought from which it can be diverted only by vigorous effort. As it is never absolutely dormant during the waking hours, it is always then either engaged more or less attentively, or allowed to follow such vague, desultory reflections as may be suggested from within, or such external impressions as may chance to captivate it. Whensoever this state of listlessness becomes habitual, the mind becomes dissipated, Thoughts are not easily collected, nor is attention readily commanded. Hence the importance of cultivating a habit of attending to whatsoever ought to be engaging the Thought, especially as the manner of using the faculty determines the measure of all attainments, material, intellectual and moral.

Thoughts are the most enduring wealth men can acquire. Neither moth nor rust can corrupt them, nor can they be taken away by force or fraud. They constitute an essential qualification for the discharge of the duties of life, and determine the happiness or misery of their possessor in a far higher degree than the external circumstances of life. They may be made the means of minimizing the ills and multiplying the blessings of life according as they are employed wisely, or are allowed to brood miserably. Of all the wealth men can accumulate, during their sojourn in the world, they are the only part that can be retained for use in a future state; the only treasure of which men will stand possessed when they render their account of the stewardship of the talents committed to them by the Almighty Father. When the earthly house shall be dissolved, and the things of time and sense shall have passed away, all that will withstand that terrible dissolution will be the work of the thinking faculty, embodying life's experience in imperishable Thought.

Their thoughts determine what men really *are*, whilst their outward acts are only what they *seem to be*. Thoughts are the evidences by which God judges, and are often very far different from the external covering exhibited to the world. The vindictive project that failed of its fulfilment; the covetous desire that could not compass its gratification; the lustful Thought that lacked its opportunity; and all the evil designs from which men refrain from cowardice, regard for appearances, or fear of consequences, are concealed from the view of their fellow-men but stand exposed to the searcher of hearts.

Thoughts are a man's constant companions. They are ever present in this life, and will continue in the world to come. As an essential part of his being, and incorporated by his own act into his very self, they are always, and sometimes very obtrusively, occupying attention. A man's character is generally estimated by the company he keeps, but a safer criterion, if it could be ascertained, would be the Thoughts he cherishes, the currents he allows to engross his secret reflections. These, therefore, are the chief factors in making or marring the happiness, and in determining the issues of life. They must be either *guests* entertained by choice, and under the control of the Will; *intruders* who have been unwisely allowed to force an entrance into the sanctuary of the heart, or they must be *casuals* who, finding the admittance free, and all within the mind careless and passive, come and go in a desultory and disorderly way. In the first-mentioned case, to use a figure of speech, the owner of the mental castle guards it vigilantly, and governs it with vigour and wisdom. He opens his gates to such candidates only as will increase his strength, and add to his resources. He sedulously excludes all such as would weaken his power. He knows that if he be not the master, he must become more or less the slave of those whom he allows to settle there. Order prevails in his ranks, and he can muster them at will. He organizes his forces as one who feels that they are available only in so far as they are obedient to his command, and own his authority. His castle represents a well-regulated mind, whence evil thoughts are excluded, because of the infection they infuse. The source from which proceed the issues of life is kept pure and unpolluted.

The owner of the castle, in the second case, carelessly admits invaders, who undermine his authority, and divide or distract his forces. Disorder is introduced into the ranks, and ruin impends. Unless the master see his danger in time, and make a vigorous effort to eject the disaffected, and restore order, the fortress must fall into the hands of the enemy. This case typifies a mind in which Thought is unguarded, control impaired, and external appearances are more studied than internal purity. Unless conscience be awakened, and roused to effort, and unless the aid of the heart's great ally be sought, a mind so divided is likely to fall into that lower domain which is of the earth, and lacks the qualification for spiritual life.

The third case is the commonest and the worst. Here the internal arrangements of the castle are disregarded, as of no concern. The gates are open to all comers, without regard to character or consequences. All are alike welcome who please the owner's fancy, offer a present gratification, or promise a future indulgence. The outside of the castle may be decorous and well-seeming, but the forces inside are motley, irregular and unruly, unequal to any great effort or serious purpose. The mind thus illustrated is one in which the important functions of Thought are not recognized, nor the condition of the inner self regarded as the real exponent of the man. The thoughts are allowed to wander and are attracted by any passing sensations—or reflections that amuse for the moment, or promise some future enjoyment. Vindictive feelings, hard thoughts, envy or discontent, are allowed free scope within, though they may be concealed from outward manifestations. Under the restraints imposed by social obligations, and regard for public opinion, the outward conduct may be respectable, or even estimable, but within is disorder and corruption.

In these and all the other innumerable types of human character the key of the actual position is in the domain of Thought, and not in the external aspects of life; in the *inner* and not in the *outer* self. Yet the mass of mankind proceed practically on the opposite principle, that, it matters little or nothing what they think, and that all depends on what they do. All men consider, at some time of their lives, that a time will surely come when their bodies, their material nature, must dissolve in death. Few can altogether evade the Thought, in that connection, of a future state; when the body, with all its appetites, sensations, pains and pleasures, shall have been resolved into its elements, and returned to fulfil some other part in the eternal cycle of cosmic evolution, yet, even amongst those who indulge definite hopes of a future state of spiritual life, few reflect on the conditions involved in such a state. How few there are, that is to say, who consider what would actually remain in their possession, when they shall have been divested of their material bodies; when the flesh, with those appetites and lusts which were, in this life, the law of their members shall have passed away, and the only part of their nature left to them will consist of the spiritual element. Whatever form may be imparted to the spiritual body, Thought must necessarily be the actuating principle, the link that connects it with the earthly life, if it be not the substance and essence of the spiritual body itself. It will be the only part of man's present self which will survive the corruption of the fleshly tenement, and the only equipment he will possess for a purely spiritual life. "When the kind hand of death shall draw aside the curtain of the flesh," what will then remain of the former self, but his Thoughts? He will then possess a spiritual body, composed of, or comprising, the spiritual elements formed and developed by the Thoughts of his heart. Then it will appear how incomparably greater is the part which Thought fulfils in the theatre of life, than even the outward actions that have represented the individual to the eyes of his fellow men; how much more important, that is, that evidence will then be by which God judges him, than that by which he had been judged by his generation.

The habit of Thought is even more subtle and powerful in its influence than other habits. To it is due, in a greater degree than to external circumstances, the tendency of Thought to gravitate in particular directions, and to determine the idiosyncrasy of individuals. It affects deeply, if it does not decide, the use or neglect of the highest faculty. On it depends the kind of control exerted by the Will, and it gives colour and tone to the opinions formed and the judgments passed, softening or darkening the natural hues of things, and imparting to them a personal or special character, not proper to themselves, but to the light in which they are viewed. The difference between the grave and the gay, the cheerful and the gloomy, the hopeful and the desponding, between the thankful and the discontented, the observant and the unheeding, is due in a far greater degree to the acquired habit of Thought than to the force of external circumstances. This is plainly proved by the different effects, observable in daily experience, of the same conditions on different individuals. One is cheerful and contented under circumstances which depress and discourage his neighbour. One sees only blemishes and defects in the things about him, whilst another sees in those same things the beauties they possess, and the purposes they serve. One regards everything in its relation to himself alone, whilst another unselfishly considers its bearing on the general weal. One, in short, having acquired by indulgence a habit of thinking which casts a shade of gloom over everything he sees ; another, of a happier mode of thought, sees the sunshine behind the clouds, even when these for the moment conceal it. Such is the practical effect of habit of Thought.

Seeing then how much depends on the use that is made of the thinking faculty, in relation to both the inner self and the outer conduct; to the happiness and usefulness of this present life, and to the prospect of the life eternal ; the natural history of the faculty, from its birth in babyhood to its maturity in manhood, becomes invested with the deepest interest. Every phase of its progress, from the earliest dawn of intelligence, and first efforts of the Will, up to the full development of both ; is full of interest, not in itself only, but more especially in its influence on the succeeding-stages. Each stage of the progress is the basis on which each successive one is formed, thus showing the vast influence of those which first occupy the virgin soil in imparting the tone to all that follow.

In succeeding chapters we endeavour to demonstrate, by an appeal to the fact of the work actually done during the very first period of babyhood, what must necessarily have been going on in the infant mind, unobserved perhaps, but as certainly as the more obvious development of the corporeal faculties effected during the same period. The keen susceptibilities and sensitive feelings of babyhood give rise to strong and enduring impressions. These and the ideas suggested by the treatment infants receive, and by the environment on which their new powers of sensation and observation are exercised, during that trying and helpless stage of their experience, constitute the aspect in which this life is first presented to them. Considering the influence of treatment and environment on the adult mind, as shown in the experience of life, it is certain that their effect must be still more influential upon the fine sensibilities of infant nature, more especially as the babe has no previous experience or knowledge wherewith to reconcile himself to present suffering. It is certain, in fact, that every part of sentient life contributes its appropriate quota of the total sum of Thought and experience. Each separate act of Thought or deed, every single experience helps to build up the edifice of the individual self, and fulfils its part, for good or for evil, in the formation of the character, and the conduct of the life. Hence the paramount importance of a right beginning, as the only safe prelude to a happy end.

The fact that little if anything has been done towards systematic training of the thinking faculty in infancy, is indisputable; but the reason is not so obvious. It seems unaccountable that the principle of beginning at the beginning, which prevails in every other part of education, should be neglected in regard to the highest of the faculties! There appear to be but two reasons to which to attribute the exception. *First*, that the infant faculty is supposed to be not susceptible of training; and *secondly*, that the process of thinking is altogether of too abstruse a nature to admit of treatment, until it is able to assert itself, and prove its competency to understand matters so recondite. Neither of these reasons has any real force, both being opposed to the facts of experience. For the progress actually made in the use of the faculty by every sane child in the first of our periods, as shown by our summary at the end of chapter vi., proves that, during that neglected period, a very considerable amount of mental work has been accomplished. Sensuous, mental, and even some moral ideas have been formulated, reflected upon, and elaborated; by comparison, association and inference. The faculty has indeed been actively exerted. In default of systematic and careful direction, it has taken such haphazard course as the casual environment may have imparted. It has been exerted upon whatsoever material the treatment of the child's tender feelings may have excited, and the attractions his surroundings may have offered for his observation. The second reason advanced amounts, in fact, to postponing the work of educating the mind until it has already been educated. Hence the laws of logic and mental philosophy attract little attention, because they are addressed to adults who have attained the maturity of their powers without their aid. They rather encumber with help those who have already helped themselves, than offer it to those in need.

We do not overlook, nor do we undervalue, the work of Kindergarten training, of which due notice is taken in the sequel, but in the meantime, whilst admitting its uses and advantages, we do not find in it all the conditions we deem requisite for infant training.

The mental condition, temper and other characteristics, acquired by unobserved mental exercise during the first periods of infancy, and which become undeniably manifest in the conduct of children a little later, are generally attributed to inherited tendencies, and to unavoidable natural depravity. Hence, though they are deplored and punished, they are regarded as inevitable consequences of a corrupt nature, and no means, or at least no adequate means, are employed to avert them. This we believe to be a cardinal defect of current practice. Our aim in the succeeding chapters is, to show that, after making all due allowance for what is really natural, the gravest faults developed in the conduct of children might be prevented; that precautions, both reasonable and practicable in themselves, would, by watchful judicious observance, avert a wrong and substitute a right direction of the thinking faculty; and that thereby a foundation would be laid by which the future character would be saved from much deformity. We shall further show that much positive evil is actually, though inadvertently, taught to infants who are thus instructed in vice, which they imbibe, imitate, and form into habit.

Referring, as is done throughout these pages, to the use of the thinking faculty as the only means by which ideas are formed and fixed in the mind; and relying, as we do, on the direction and cultivation of that faculty for all the power attainable, and for all the knowledge men have any hope of possessing, we must not be supposed to overlook, or to underrate the office and work of the Holy Spirit, nor yet of prayer, as the most natural, and at the same time most exalted, use of the faculty itself. But, in recognizing the inestimable resource which a true Christian possesses, in his appeals to Supreme Power, in his weakness and extremity, we do not impair but strengthen the position we assume. For Divine aid is given only to the willing and the weak, and must itself be sought. Spiritual aid supplements but does not supersede or overrule the individual faculties and Will. Neither Thoughts themselves nor the Will by which they are prompted, are forced upon men, who know too well, indeed, that they may and do resist the most powerful appeals from within or without, however impressively they may be urged on unwilling minds. The fact that men may invoke a higher power to aid their real infirmities, affords no excuse whatever for neglecting the use of the powers they possess; on the contrary, such help being supplementary, it can only be claimed after means have been fully exhausted and the use of the faculties have been energetically applied.

Introduced to a life in which good and evil are inextricably intermingled, a life which is, in fact, given expressly to enable men to discriminate and choose between these conflicting principles, the training of the novice, who is entering on the conflict thus imposed upon him, should be conducted primarily with a view to the formation of correct ideas of moral principles, as the groundwork on which the edifice of Thought, that is, of the individual self, is to be built. By this we do not mean forms of belief, or strictly religious instruction and doctrine—which are particular applications for after consideration—but, as fundamental elements of character and conduct, equally important in regard to secular knowledge and affairs, as in the more particular sense of religious doctrine. Good and evil are so intimately and universally implicated in *all* Thoughts, opinions, judgments and general conduct, and do so closely concern the inner, as well as the outer, man, that their influence cannot be excluded from consideration in any part of either knowledge or conduct. The sincerity and the force with which the conflict between the higher and the lower law is waged throughout the course of life, depend so much upon faith in the authority which so continually imposes restraints upon the natural appetites and passions, that the ideas of faith in tutors, and voluntary obedience to their authority, as first inspired by their example in the earliest stages of infant life, are the most influential elements ever acquired for the after-development of character. As means of initiating and promoting discipline, these elements are of the utmost value, but their chief function is in preparing the way for, and establishing a right understanding of, the higher principles of faith in, and obedience to, Divine authority. Here, as throughout the history of life, that which is natural precedes and prepares the way for that which is spiritual.

If we could believe that the defects of current practice in the training, or rather in the neglect, of the education of the faculty of thinking in early life were simply one of the many ways in which men fail to exemplify in practice what they nevertheless know and believe theoretically, we should relegate the task of the practical enforcement of known duty to the pulpit, whence such admonition would more appropriately proceed. Long and varied experience, and close attention to this particular subject, however, satisfy us that the worst results of current practice are referable to causes, the force of which is not generally recognized. For the purpose of establishing this position we have endeavoured in the sequel to trace the successive stages of the progress and development of the faculty in a natural history, beginning at the beginning and proceeding to that stage of the understanding where it emerges from the discipline of tutors and sets out on the independent career which is to complete the structure of the self.

For convenience of treatment, we have divided the history into periods. These are necessarily of an arbitrary nature, and are not defined by any clear lines of demarcation. The process of development being quite continuous, and very gradual, does not exhibit any such sharp lines as would serve to distinguish them very clearly from each other. Language, however, is so influential an agency in the development of the thinking faculties, and in the intercourse of men, that it serves better than any other means at command for regarding separately periods or stages of early life which could not be conveniently discussed collectively.

Such being the general scope and aim of our undertaking, the following chapters are devoted to a systematic discussion of the rise and progress of the faculty of thinking, and of its practical use, divested as much as possible of its mysterious and metaphysical characteristics. These, though alluring and seductive in their scientific interest, have no actual bearing on the use of the understanding. Nor do we profess to treat the subject technically or with rigorous scientific exactness. We do not question the interest attaching to pure mental philosophy and the science of logic, but they do not affect our purpose. The noblest characters that have figured in history were formed without any reference to these abstract considerations, which in fact do not present themselves to notice in the affairs of life. The greatest achievements of the intellect, in art, science, religion, and in the most advanced forms of knowledge have been accomplished without their aid by men who had no better knowledge of the mystery of Thought than the humblest of their contemporaries.

Thought, as the product of the thinking faculty, the result of a process, will be considered as distinct from the faculty by means of which it is originally formulated and afterwards elaborated by reflection, comparison, association, and inference. The Will, as the actuating and controlling power, will also be separately discussed. These factors must be considered separately, if either their individual or joint action is to be rightly understood.

Whether perception and reflection are separate powers, or are different functions of the same faculty, is immaterial to any part of our discussion, as is also the question whether or not the Will is a distinct power, or only a different function of the principal faculty. The fact that involuntary action exists in the highest types of animated beings associated with voluntary power, suggests the probability that the Will is a special faculty. And this idea seems to receive further sanction from the fact that, in descending the scale of being from the higher to the lowest types, there is a point where voluntary action ceases, and its function is fulfilled by some affinity or mutual attraction existing between the animal and its proper food : such as exists in plants.

Whatever may be the essential relation of the Will to the other functions or faculties, we discuss it separately, because of the influence it exerts in the conduct of life, and because of its early assertion in the period of babyhood. The treatment of the Will in the very earliest indications is, in our belief, the most effective means of establishing in the young mind that voluntary subordination to moral suasion which constitutes genuine obedience. The same considerate treatment will go as far as human means can go towards the formation of a docile and amiable temper. One of the principal uses of the Will in mental operations is to secure that control and concentration of the powers which is understood by the term *attention*. This quality, so valuable to both scholar and teacher, yet so difficult for children to practise, cannot be forced. It must be *won;* and the most efficacious means to that end is by the early education of the Will, the dominant power of the mental estate. This subject is discussed in the chapter on *Control* of the Faculty of Thinking.

Throughout this work the term "tutor" is used to denote parents, nurses and elders generally, who exercise authority, and it comprises them all, so as to save needless verbiage. Where a teacher, or instructor, is meant, one of these words is employed, in order that the term *tutor* may be reserved for the more general purpose. The male sex is, in accordance with precedent, used to comprise both sexes.

CHAPTER II.

ON THE NATURE OF THOUGHT.

Of all the subjects that could engage the attention of man, there is not one more worthy of it, nor any that would better repay it, than that of Thought, the product of his highest faculty. There is no other so important to him in the conduct of his life, whether regarded in its relation to this world or the world to come. Thought embodies all that is enduring of his experience of life and the lessons it teaches, all the knowledge he has acquired, and the only hope of acquiring more. There is therefore no other subject that concerns him so deeply, nor any of which a close study would reward him so richly. The substance of life's circumstance is summed up in Thought, and in that form alone can it survive the dissolution of the material world.

Thought, in its collective sense, the accumulated body of Thought, which comprises the result of all observation, and the work of reflection and imagination, is composed of ideas, or individual Thoughts, each of which is the product of the formulative function of the thinking faculty. This function is exercised, generally, by an effort of the Will, on such of the perceptions as engage attention. A large proportion of all that passes within the range of perception, however, escapes the notice of even the most observant, or is intentionally disregarded. Such sensations as are altogether neglected, and receive no attention at all, are wholly lost, as they are not formulated into ideas, and therefore are not gathered into the storehouse of the mental possessions. But, between the two extremes of total disregard, on the one hand, and the strongest effort of attention on the other, there may exist every possible degree of notice. Hence, the ideas which compose the stock of Thought vary in strength of impression and clearness of definition in every conceivable degree. Attention, therefore, whether it be prompted by a resolute purpose, or be captivated by the irresistible force of the sensation or reflection, is the chief factor in determining the fixity and value of the work of the thinking faculty. This important factor will be more fully considered in the sequel, in the chapters on *Control* and *Memory*, in the discussion of which subjects its bearing is most conspicuously manifest. At present it is noticed only in its bearing on the natural history of Thought, as that on which depends the depth and clearness of the ideas resulting from the formulative process. These qualities determine the nature and value of the collective body of Thought and the extent to which its elements are amenable to recall by the memory, for the purposes of reflection, and for use in the conduct of life.

Hence it follows, that the sum of Thought, the stock of ideas acquired during life, does not represent the opportunities life has afforded, but only the use that has been made of them. It consists, not of the things submitted to observation, but of those only which, by vigilant and well-directed use of our highest talent, have been rescued from amongst the fugitive scenes of life's panorama. The records inscribed on the mental tablet, even of those few which have been so turned to account and appropriated, are characterized, unfortunately, by such error, defect and distortion as are incident to the infirmity of the instruments employed in noting and collecting them.

Every idea, or individual Thought which enters into the collective stock, is the result of a process, in which several factors are involved, viz. :

1. The subject-matter.
2. The perception thereof, whether by sensation or reflection.
3. The Will, whether to notice or to neglect the perception.
4. The formulation of the subject-matter into an idea and its incorporation into the mind.

Each of these factors is subject to influences which vary in every particular case. The *subject-matter* may be important or trivial, simple or complex, and its presentment may be clear or obscure, faint or forcible. The *perception* may be vivid or dull, complete or imperfect, transient or enduring. The effort of the *Will* may vary in every degree, between intentional disregard and a vigorous effort of attention, or it may be forcibly attracted by the pressing nature of the subject-matter. In these and many other ways may the essential elements of the process affect the final result. The bearings of these influences on the quality of the ideas produced are more fully discussed in the sequel, in the chapter on *Memory*, as the amenableness of an idea to be recalled is the crucial test by which its practical value is estimated. It stands to reason that an idea which cannot be recalled, which is not subject to the action of the memory, can neither be reflected upon nor used. If the record be illegible, it is of no practical avail. Whatsoever qualities, therefore, are necessary to insure a prompt response to the summons of the memory, they constitute the real value of every idea that enters into the treasury of Thought. A thorough knowledge of these qualities, and of the best means of acquiring them, is of paramount importance in enabling us to turn to account the priceless opportunities which this fast fleeting life affords.

The sources whence the subject-matter of thought is derived are *sensation* and *reflection*.

Sensations proceed from (1) internal feelings and appetites, which are mainly involuntary; and (2) external objects and circumstances of observation.

Reflections are, in their nature, exercised upon ideas previously existing in the mind, and comprise the several processes of reasoning, inference and imagination, whereby new ideas of the relations and properties of things are acquired. Ideas are thus multiplied, and knowledge is extended beyond the range of mere observation.

The subject-matter of Thought is primarily supplied through the medium of the senses, and is at first suggested entirely by corporeal feelings and appetites. Hence the first ideas are of a sensuous or animal nature. As the organs of sense come into more extended use, ideas of external objects begin to be acquired, and afford subject for the exercise of the reflective faculty. As environment enlarges, and the *mental* faculties, as well as the organs of sense, become more amenable to the Will, the power and range of perception increase, and the subject-matter on which the thinking faculty is exercised embraces a larger and ever-extending sphere of observation and reflection. External circumstances, as well as external objects, shortly come within the sphere of the perceptions, and foremost of these is the conduct of tutors, especially as it is exhibited in their treatment of the infant himself. This forms an influential subject of his early reflections, comprising the first suggestions of the *moral* element of Thought, the first ideas of good and evil.

The Will has no part or influence in regard to the subject-matter on which the faculties are exercised in the earliest stages of infancy, as corporeal feelings which form a necessary part of the conditions of the life of the young creature are involuntary. His attention is not long entirely so monopolized, however, but is attracted by external objects and circumstances, and his mental faculties begin to be exercised in reflections thereupon. Thus begin his relations with the outer world, which is for the present comprised within the narrow bounds of nursery environment. From this sphere he will eventually emerge, with such relations as may there have been established in his mind. All his future relations to the outer world will be founded on those first formed, and will thence derive their tone and character. The subject-matter of future Thought will be increasingly, though slowly, brought within his own choice and control, until he shall have attained to such a degree of independence as will constitute him a responsible free agent. What will be the nature of the choice he may then make, and what the extent of the control of his thinking faculties he may have acquired during his period of tutelage, will depend on the training his thinking faculties may then have received—whether this training have been effected by the chance influences of his surroundings, or by the systematic efforts of tutors. The direction or bent the Will may have taken during the earliest stages of youth will determine to a very great extent whether his aspirations and pursuits partake of the lower or higher class of Thought—whether the sensuous, mental, or moral elements will prevail in the formation of his character.

Formulation is that factor of the process by which an idea is inscribed, as it were, on the mental tablet, and which gives it the form in which it is incorporated into the mind. It is effected under very various conditions of the faculty itself, of the presentment and perception of the subject-matter, and of the attention given to it. A great many of the impressions on the sensorium are lost for want of the effective fulfilment of this essential part of the process. Even of those objects which are actually perceived, comparatively few are so formulated as to enter into the available stock of ideas. The fact of their having been perceived is not unfrequently attested by responsive acts done in a mechanical way, unconsciously. A good illustration of this kind of perception and response, and of the utter absence, nevertheless, of any formulated idea of the circumstance, is given in Note 7 of the Appendix. Such experiences are of daily occurrence, and this one offers no very remarkable features, except in its being so plain, and the demonstration so complete, as to the failure of the formulative action. In a very preoccupied state of mind, sensations even of a striking character may fail to arouse the faculty whose proper function it is to formulate and appropriate them. The familiar voice of a friend, or even peals of thunder, may be lost to the mind, although they must have reached the sensorium. The same lapse may occur from habitual indifference, or what is termed an absent state of mind. Of all the objects seen in a single hour's walk, a very large proportion are altogether unnoticed, including some which would have well repaid attention. Persons of observant habit thus reap a rich harvest of Thought which is lost to those of an opposite character.

Nor is it only in regard to objects submitted to the senses that the formulative function is necessary. Reflections suggested from within may be perceived with so little regard that no proper apprehension of them ensues, but only a superficial sort of impression is produced, not such as to evoke a clear effective formulation. The perception in such cases is weak, not for want of natural power, but from inattention. Minds of this habit wander ineffectually and may altogether lose the power of concentration, for want of the habitual exercise of attention. Persons so affected complain of their defective memories, and their inability to collect their thoughts. Being generally unconscious of the cause of these defects, and that the fault rests with themselves, they would feel deeply hurt at any imputation of defective effort. They regard the superior powers of their neighbours in the same way that indolent persons regard the fruits of industry, and attribute them to the favour of Providence.

From the foregoing remarks on the part which each of the factors has in the resulting idea, it follows that the ideas which compose the collective stock are of widely different force and value, according to the conditions under which they were elaborated. Besides the difference of depth, fixity, and clearness, so produced, however, ideas are of different kinds, and may, for our purpose, be classified under three heads, viz. : (1) simple, (2) complex, and (3) indeterminate.

Simple ideas are those in which a single object, quality or relation is represented, and may consist of an image, as a star, a circle, a colour, &c. &c., or may be a perception of a sound, a pain, or a taste. Simple ideas are elementary, both in their nature and use. They fulfil the same office for Thought that letters of the alphabet do for language. Like them, they are not so much employed singly as they are in groups. Their chief use is as component parts of more complex and comprehensive thoughts. The simplest objects presented to the mind or sense are rarely regarded by the adult in their simple form, but generally in connection with some shape, quality or relation proper thereto. In regarding a cloud, for example, the idea of its simple nature as a cloud is associated with its form, colour, distance, portent, &c., and is apprehended and formulated along with its qualities, relations and significance in one complex idea.

The process of thinking is in this respect analogous to that of reading. The facility of apprehension acquired by practice makes the percipient impatient of the slow and tedious process of spelling out as it were, separately, the simple component elements, and he grasps therefore as many of them in one single effort of apprehension as may be obvious, just in the same manner as an expert reader regards not the letters singly but the words or phrases entire. An expert chess-player sees the board, the pieces and their relations to each other collectively, in one comprehensive grasp, whilst the novice has to regard the pieces and their mutual bearings separately. So also in the formulation of ideas by the infant faculty; the complex is first acquired by means of the simple ideas spelt, as it were, singly.

Complex ideas are those which comprise in one concrete form several simple component ideas. They bear the same relation to these that words or phrases do to the letters which compose them.

Language is a very convenient vehicle for expressing complex ideas, because a very extensive and complex group of ideas may often be thus conveyed in a single word. It must not, however, be supposed that the complex idea was originally so formulated in the mind. The general functions of language are discussed elsewhere in chapter iv., but we must in this place point out that, to a great extent, at least, the apprehension and formulation of ideas is quite independent of the use of this powerful agent of expressing ideas so formed. This fact is clearly proved by experience. The recognition of expression in the features of tutors in early babyhood shows that ideas of emotion exist in the baby mind, associated with the images of the features themselves. The ideas and inferences involved in such recognition, made before language has been either comprehended or used, conclusively proves that ideas are independent of that element in the original formulation. A further proof, deduced from the action of the adult mind, exists in the difficulty frequently experienced by highly intelligent persons to express, that is, to clothe in language, ideas which exist in their minds in a definite form. It is not the function of words to formulate ideas in the mind, but on the other hand the mind devises words to convey the ideas already existing there. Words in fact are of no use except in so far as their meanings, that is, the ideas they express, already exist in the mind. The meanings of words are derived by inference from the circumstances of their *use*. Hence, if the inference have been erroneous or imperfect, the *word* will be wrongly used. The word represents the idea, and becomes associated with it.

Both simple and complex ideas are determinate. The former are so in their nature, and the latter as being resolvable into their simple elements. This, however, is not the case with a very large and important class of ideas which are indeterminate and not so resolvable, or, if at all, only by much circumlocution, and often even then with very doubtful result.

Indeterminate ideas differ from complex ideas in not being resolvable into simple elements. They are generalizations, of which the particulars are too numerous, indefinite, or obscure, to have been formulated in detail. They are of the nature of general inference or insight, and may take the form of prejudice, attraction, repulsion, surmise, and other such impressions. They are verdicts conceived by the imagination, in which they differ from judgments formed on recognized data of evidence.

Notwithstanding their indeterminate nature, they constitute a very considerable and highly influential part in determining character and conduct, especially in regard to the relations of the individual with the outer world. Though they are generally the result of more or less extended intercourse or observation, they are sometimes wrought by a sudden and inexplicable impulse. Of this latter class, a familiar example is given in the saying: "I do not like thee, Dr. Fell, The reason why I cannot tell, ... I do not like thee, Dr. Fell." Of the former, the word *truth*, so commonly used, and so well understood, may serve as an illustration. The idea represented by the term is so familiar as to become almost determinate. Any analysis would seem superfluous, yet the idea is derived from many experiences, and in most adults undergoes some considerable change of significance from that it possessed when it first was formulated, and used as an idea in childhood. The term "gentleman" is almost as commonly used, and as little likely to be misunderstood. Yet two persons who would apply it with equal propriety, would probably describe the conditions implied by it in very different ways, and not without considerable embarrassment. Still, the word represents a distinct idea, and as an element of knowledge, it is one, not several; an idea in the proper sense of the term. Indeterminate ideas, such as these just mentioned, which are in current use, necessary elements of common conversation, are reduced almost to the practical simplicity of determinate ideas by the aid of language. Some others, which are equally common, are not understood so well, and are employed in senses differing according to the various ways in which they exist in different minds. Hence arises not a little misunderstanding, especially in argumentative discourse, wherein each person uses the term in the sense in which it exists in his own mind.

Those indeterminate ideas which most concern individual conduct, are of a more special and limited nature. Not being in common use, nor of more than individual interest, they are not expressible in single words or phrases by a common understanding. Hence they are altogether independent of language, were not formulated in such terms, nor so exist in the mind. Nevertheless they are strong, enduring, and obey the summons of memory in their integrity. Treatment that has been received, a sermon heard, or a book read the details of which, in each case, offered no salient points to invoke the formulative function, may nevertheless make an impression, long and definitely remembered, and such as to affect materially the conduct.

We lay stress on this kind of idea, because of its bearing on the natural history of infant Thought, of which a great portion is gleaned from the general experiences of that trying period, and is thus indeterminately summarized in the form of disposition, temper, and a general relation or attitude to external persons and things.

In this work, which does not profess to deal with mental philosophy technically, but only in its practical bearing, we might well omit any notice of the doctrine of innate ideas, which was long since discarded, were it not that, under the guise of special gifts, it practically survives. These are discussed in our chapter on Inheritance, but we deem it proper in this place, as a preliminary to the discussion of the first beginnings of Thought, and its future development, to maintain that all the knowledge man possesses, or possibly can acquire, is obtained by the use of his thinking faculty, and by no other means. We do not thereby exclude the work of the Holy Spirit, as that also comes in response to the efforts of the faculties; nor do we question the existence of great diversity of natural power of the faculties, as that is taught in the parable of the stewardship of the talents five, two and one, respectively. We fully admit great difference of power in the thinking faculties of different individuals, but we deny their endowment with ready-made Thought. The instruments are a natural endowment, but the work is done by the possessor of them. Were it possible, however, that ideas could be conferred, and that in addition to the faculty for acquiring knowledge, we should have also a treasury ready furnished, ideas so gratuitously provided would be of no use. They would be like words in an unknown tongue, without use or meaning. The significance of an idea, even its bare intelligibility, depends on the circumstances of the presentment of its subject-matter. It is from these it derives its meaning and value. The doctrine is untenable, and is as unreasonable as that of the worthy bishop, mentioned in chapter xiv., who held that fossils were created ready-made, and therefore afforded no proof of the pre-Adamite existence of the world.

The theory is unsupported by positive evidence, and its assumption is superfluous, as all the practical phenomena of Thought are accounted for by the use of the ordinary means men possess for acquiring knowledge.

The thinking faculty, though it is the highest man possesses, is not monopolized by the human family, but is shared also by the lower animals, in such measure and manner as is necessary for obtaining their food, eluding their enemies, and for other purposes proper to their mode of life and ultimate destiny. It is not the possession but the use of the faculty which distinguishes the human from the brute races. Hence, it is to be considered in its relation generally to all beings possessed of any measure of intelligence. A natural history of Thought which should exclude its relation to the lower animals would be incomplete, if not misleading.

Whereinsoever the use of a thinking faculty is common to both the lower and the higher races, much instruction is to be gathered from a study of its operation in the lower, as regards its use and functions in the higher, domain of Thought. In the lower types it is divested of some conditions which complicate and embarrass the consideration of its use in the higher. The bearing of Thought in its higher development receives effective elucidation from its operation in the lower, especially as the personal history of the human subject commences with the lower and passes through all the intermediate phases of condition up to the higher. Beginning life with animal faculties only, the progress of the individual proceeds through the successive stages of sensuous, mental and moral nature, each in turn exercising paramount influence, fulfilling its special part, and impressing its proper stamp upon the constituent Thought.

In succeeding chapters we avail ourselves of illustrations and examples, adduced from the operation of the thinking powers of the lower animals, which afford convincing and instructive evidence in regard to the progress and development of the faculties of the human subject.

If the reference to brute nature needed, under these circumstances, any apology in a work on the practical use and bearing of Thought on human conduct, it would surely be found in the fact that the whole creation, animate and inanimate, owes its being and the order of its existence to one Almighty Father, whose creative power, supreme government, and paternal care, are alike manifest in all the multifarious elements that constitute the cosmos. No one of these can say to another, "I have no need of you," each is essential to all the rest, and takes its appropriate part in the drama of Nature. We do not understand or sympathize with the feeling which has been suggested to us for the omission of all reference to the lower types of intelligence, as though such allusion tended to lower the dignity of man's place in the Universe. We consider, in fact, that the superiority of man's position in the scale of being was established by the Almighty fiat which conferred upon him the dominion over the whole of the brute creation, and thus defined the mental relations of the two orders of sentient beings. It is not a dignified position for man, the lord of creation, wilfully to reject some of the most significant pages from that book of Nature in which it has pleased the Almighty to reveal His eternal power and godhead, lest they should wound his pride. Nothing is to be gained, but much loss suffered, by self-imposed blindness.

Considered in the phases in which it is exhibited in the whole range of intelligent being, Thought is naturally divided into two distinct departments, which are sharply defined. The *lower* comprises all that is possessed by the brute races. The laws thereby imposed define their powers of Thought, the purposes to which they may be applied, and the obligations proper to the part they fulfil in the economy of Nature. The *higher* is the special endowment of man, whose powers range throughout the whole domain of Thought, including the lower, to whose laws he is also amenable, as well as to a higher and rival law, imposed on him by his higher nature. This higher power, by means of which he is authorized and commissioned to exercise dominion over the lower order of beings, imposes on him the obligation to bring into subjection also the lower part of his own individual nature. Hence arises a conflict between the law of the lower and that of the higher domain of his complex nature. The issue of this conflict determines his place and rank in the scale of being. It either subjects him to his lower nature, or initiates a spiritual life.

The researches of science prove that beings of the human form and corporeal structure existed before the epoch of the Mosaic narrative, and consequently anterior to the creation of the image of God in man. That likeness could not have existed in the outward and corporeal frame, for God has not limbs, or organic structure. It was, in fact, in his knowledge of good and evil, as is implied in the passage: "Behold the man has become as one of us to know good and evil." The *moral* function of Thought, as distinguished from the lower domain of sensuous and mental Thought, begins with the knowledge of good and evil, the basis of moral principles. These constitute the higher law of spiritual authority, by which the lower is to be subjected; and there is a stage in the conflict between these two laws, which is a condition of the second birth, the introduction to spiritual life.

The second birth of which our Saviour spoke to Nicodemus, as being the introduction to the spiritual life, was not then originated as a new doctrine, but was an old truth clothed in a particular formula of words. He could not have intended to imply that the patriarchs, and other exemplary characters of Scripture, had all been wanting in that, or any necessary qualifications for the enjoyment of spiritual life. Some of them had such communion with the Almighty as proved their having attained a high position in that life. By the second birth, therefore, or entrance into spiritual life, we are to understand not only a *knowledge* of good and evil, but an effective choice, a firm and faithful embracing of the good; a stage of that conflict between the lower and the higher laws of our nature, which indicates the triumph of the latter, when evil has been effectually overcome by its natural counter-agent, good. Nor is this doctrine peculiar to the Christian religion. The conquest of appetite and passion, that is, of the brute nature, is professedly the one great end and aim of Buddha's discipline. According to his doctrine, the complete subjugation of the lower nature of the individual entitles him, thus freed from appetite and passion, to be absorbed into Buddha. This state is often misunderstood to mean annihilation. Such a climax, however, would offer but an ineffective stimulus to any reasonable aspirant to undergo the mortifications and trials imposed by the stern régime of Buddhism. But the state, as described by Buddha, in several parts of the Buddhistic writings, is one of peaceful bliss,* and therefore far removed from annihilation, and much more resembling a spiritual ascendancy of the higher over the lower nature.

The spiritual touches the material world in that higher domain of Thought of which the knowledge of good and evil forms the basis, and the triumph of the former a primary qualification. Man, therefore, possessing such knowledge, has the key to the attainment of spiritual life, and he who, in the exercise of his free will, elects to embrace it, will not want either the gift itself or the power to enjoy it. Animated by such aspiration, evil would in its nature be repugnant to the candidate. Repentance for his past lapses would be natural, deep and genuine, just in proportion as his choice was sincere and earnest. Repentance, therefore, and faith are natural conditions of the act of choice, for he who so acts must necessarily believe firmly in the doctrine and in the authority on which it rests, that of our Blessed Redeemer who sealed it with his life's blood. It follows, therefore, that faith and repentance, the conditions prescribed by the gospel, are not mere dogmas, but natural operations of the higher law. The conditions necessary for a spiritual life being present, that gift will represent the second birth, and thenceforth the natural man is endowed with a spiritual nature, and will enjoy communion with the supreme spirit, more or less closely, according to the strength of his faith, the earnestness of his purpose, and the effort made for its fulfilment.

* Rhys Davids in his work on Buddhism gives a list of these passages.

In contemplating the visible universe, generally, the evidences of a power superior to the forces therein operating, and to the laws by which they act, must be recognized. The changes that are continually in progress by means of those forces, and the resultant developments therein manifest, attest the existence of order and purpose, and consequently of a power exercising supreme control. The Being who wields such supreme power must necessarily be of a different nature from the material subjects of His rule, animate and inanimate. In Him, therefore, is revealed another order of being, which typifies the spiritual element. The material and spiritual natures both stand revealed in the order of the universe, and both are apprehended by the agency of the thinking faculty in the exercise of its highest functions.

In the lowest orders of the scale of living beings the vital principle only is evident. Even sensation seems wanting. Vitality in them seems to be maintained by some kind of mutual affinity, which subsists between them and the fuel they consume. Sensation is not necessary in their condition. This faculty appears very low in the scale, however, and volition soon follows. These graduate through a long series of successive orders, in which the rudiments of mental processes become manifest, and these eventually become developed into the familiar forms in which they are exhibited by the more useful and companionable of our domestic favourites. Thus, from the simply vital to the advanced mental conditions the lower order of beings progress, and at this point their further progress stops. A similar order prevails in the incipient stages of the human infant, and hence the interest the natural history of the lower animals possesses in contemplating the development of the lower element of the human mind. Here, however, the further progress is not arrested as in the brute, but proceeds upwards from the lower into the higher domain, which may culminate in a spiritual life.

In the following diagram the course of the development of the thinking faculty in the economy of the animal kingdom is briefly sketched in a summarized form, as a preliminary to the discussion of the rise and progress of Thought in the human subject.

DIAGRAM.

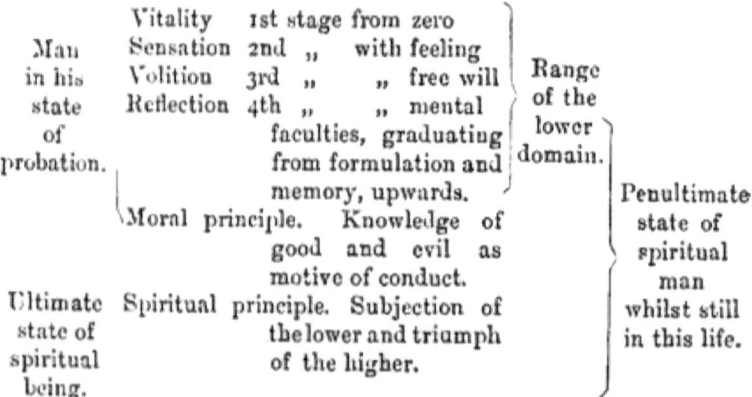

The purpose of the foregoing reflections on the two departments of Thought, which, though so distinct in their nature and objects, are so inextricably mixed together in this life, is to show their practical bearings on the cultivation of the intellect, and to prove the necessity of commencing the training of the faculty from the very earliest period, when the lower nature predominates; and thence, from the basis so laid, to follow its development upwards, always in the right direction, that is, towards the higher domain of Thought.

Considering how great is the function of Thought in all the affairs of human life, as the spring of conduct, the embodiment of all knowledge and the basis of principles and opinions, and seeing that the nature as well as the extent of Thought is so influential in its practical use, there needs no further argument to prove the responsibility involved in the training and exercise of the faculty in such manner as to give a right direction from the very first dawn of intelligence onwards. So commenced, each successive stage would rest upon a sure foundation, and life would begin under the auspices most favourable to a high intellectual and moral qualification for whatever sphere the individual might occupy. A happy relation with the world in general, loyal submission to authority, a power of attention and concentration, effective memory and all the requisites of sound judgment are qualities attainable, in a greater or less degree, by every one possessed of a sane mind, provided the faculty be trained accordingly, and their opposites are unhappily the alternatives to which the neglect of such training will probably lead. The real character of the individual will be moulded to the pattern of its fundamental element, Thought.

CHAPTER III.

ON THE FUNCTIONS OF THOUGHT.

THE primary functions of Thought are to sustain the life and regulate the conduct of its possessor, whether he belong to the lower order of sentient beings or to any of those intermediate stages of intelligence upwards to man himself. In all stages of the corporeal development the thinking faculty is the agency by means of which all the purposes of life are directed and fulfilled. Each successive stage in the development of the organism enlarges the range and increases the functions of Thought, wherefore in man they attain their highest form and largest sphere of operation.

The higher function is the reproduction of Thought. Knowledge is not bounded by the ideas formulated by the direct operation of the faculty, but is extended by means of the reflex processes of reasoning and inference. By these reflective processes knowledge is extended far beyond the limits of those elementary ideas, in an ever-expanding ratio. The sphere of Thought thus embraces all creation. Overleaping the boundaries of the material universe, and all limitations of time and space, it stretches into and penetrates the spiritual world. All Nature is subjected to the powers of its research. The earth yields up the secrets of its history, growth and destiny, reveals the laws which have regulated its formation out of chaos, unfolds its complicated evolutions, and makes manifest to the keen scrutiny of the thinking faculty the eternal power and godhead of a Supreme Author.

The functions of Thought may be considered in their relations to the animated part of creation generally, or, in a more restricted sense, to a particular era in the history of the world, a nation, or an individual. In each of these spheres they operate similarly according to laws which determine the position of the age, people or individual in the scale of progress. Successive developments of material organism are accompanied by corresponding increase of the sphere of Thought, which progresses, therefore, in the collective body of sentient beings, as well as in the individual, according to the development of the organism. In other words, Thought advances in proportion to the progressive development of the organs through which its correspondence with the material world is carried on.

Considered with reference to the human individual, the functions of Thought are exercised variously, according to the different conditions incident to each successive stage of his life. These stages maintain the same order in each individual as that in which they stand in the collective body of animated beings, as sketched in the diagram in chapter ii.

Such being the natural order of Thought, according to the appointment of the Almighty Father, such is the order in which it should be studied by man, His creature. The relation in which man stands to the lower orders of sentient beings, by the operation of His beneficent laws, could not possibly imply any degradation, as all the others have been subjected to him, nor, in such relation, could it be hurtful to the dignity of man to study Thought in its whole history, from its inception in the lowest, upwards to its maturity in the highest, of its developments. Viewed from man's exalted standpoint, every successive step in the upward progress of Thought, from its earliest dawn to the highest point attained in his own order, is invested with special interest, as the heir of the one preceding, and the parent of all that follow.

Seeing also that in the history of each individual the same order prevails, and that every man must necessarily pass through every stage of being from birth to manhood, the study of the functions of Thought from its beginnings affords an insight into their purpose and progress which cannot in any other way be acquired. The interest and importance of such a complete historic consideration of the subject is intensified by the fact that, in passing through each stage, he retains in his nature its characteristics as well as its fruits. Each of the fields he has successively occupied, and sown with good and evil seed, remains as a permanent and essential part of the territory of his individual estate, and the accumulated harvests thence reaped will remain, whatever their nature and extent may be, to constitute the treasure of Thought for his future capital in the conduct of life.

The animal or sensuous, the mental and the moral elements of his complex nature, which, in turn, engrossed his faculties, remain fast bound together, along with their products, for the natural term of his earthly life. He cannot divest himself of any part of them. Each fulfils an appropriate and necessary rôle in the formation of the self, and each entails its own special responsibilities, inseparably attached to its contributions. There exists amongst them a mutual interdependence and co-operation of Divine appointment which can neither be dispensed with nor destroyed. It behoves man, therefore, to consider closely the bearings of each upon the other, separately, and on the whole, collectively. The functions of Thought are paramount in each, and confer upon it all the value it can possess as part of the whole being. It is therefore necessary to consider the part those functions perform in the lower, in order to obtain a right and complete understanding of the phenomena and purposes of their operation in the higher domain.

In man, with his highly developed organism and complex nature, the functions of Thought are fulfilled under conditions which make it difficult, often impossible, to trace their bearings on any particular part of his nature. The fact that all the elements subsist and operate together, obscures the separate influence of each. In babyhood, for instance, the vicarious offices of tutors, who are possessed of a higher order of thought, suggest ideas which are not proper to the babe's own stage of being, and thus complicate matters. In the lower animals the functions of Thought may be seen in a simpler form, being there divested of the complications introduced into the higher nature by moral considerations. These are most influential in human Thought in its advanced stages, and are thus introduced vicariously into the lower stages of baby life. Hence the natural workings of Thought in its lower domain are best understood by the study of them as they are exhibited in creatures who are born, nurtured, and matured in simpler conditions of life, free from the influence of the moral element. The illustrations adduced in the following chapters from the lower orders of animal nature will bear out this conclusion.

There is a unity of plan, a continuity of process, and a completeness in the works of Nature that cannot be fully apprehended without a thorough investigation of their rudiments. Science testifies to the fact that the clearest light thrown upon the higher branches of science has, in every case, been obtained by means of the lower. There is, therefore, no rule of science so universal as that of beginning at the bottom. The fact that such study as has been devoted to the training of Thought has been an exception to this rule is noticed in a former chapter, and referred to its probable causes; but the consequences are pertinent to our present argument. They are manifest in the popular idea that the tempers, dispositions and decided bent, so often exhibited in childhood, are inherited properties and therefore inevitable and independent of the child's own personal experience, whereas they may be proved to be the natural results of the treatment and circumstances of early infancy, when the operations of the mind are obscure and unobserved. They are, however, neither less natural nor less influential on that account, and are therefore deserving of the utmost attention.

When the human subject first enters upon life he begins it under the same conditions as the lowest in the scale of animated being. Such in the order of Providence is the first stage through which he has to pass preparatory to his realization of the high position he is ultimately to occupy in the scale of being. His sensations are those of feeling and appetite. These form for a time the sole subject-matter in which his incipient faculties are exercised. And they *are* so exercised, with results which form the basis of their future and wider operations. The first fruits of the thinking faculty are suggested by sensations of his animal nature; those very same feelings with which he will, in one form or another, have to contend through the future stages of his life. The faculty which for the present is exercised upon and dominated by them is that by which he is hereafter to bring them into subjection or to forfeit his hopes of a spiritual life. In those members which now absorb his attention is written the law which will through life oppose the higher law of the nature he is soon to enjoy. The first impulses which quicken into effort his incipient Will are those of this lower law. In his present state, however, he is not left to his own resources. The inexperience and helplessness which place him in the power of tutors, impose upon them the duty of acting for him, and also the responsibility of the issue. It is for them to satisfy his appetites and supply his wants in such manner as to modify their power, and vicariously to do for him what he may thus gradually learn to do for himself. This lesson may come slowly, but he will have ample time to learn it, and will not fail if it be judiciously taught by consistent and thoughtful treatment. The same considerate regard for his tender feelings will win his confidence and love, and enable tutors to subdue his incipient Will by unfelt restraints, and inspire his spontaneous obedience.

Thus is commenced by proxy the suggestion of that discrimination between good and evil which engages the highest function of the faculty in after-life, and thus also is inspired the choice of the former. Hence, though the Will is in both orders first prompted, and brought into exercise, by appetite and feeling, the ultimate subjection of the lower nature in the one case, and its lasting predominance in the other, are plainly contemplated by the order of Providence in the respective conditions of life in each. In the lower order, the Will is subject only to a rough discipline of force, but is otherwise unrestrained, and left to develop the natural propensities proper to the class of animal in each case. The brute cultivates his lower nature, which is necessary to the mode of life he has to follow. The human subject has these same propensities to subdue. In the former, ferocity and self-assertion are virtues. In the latter, they are the forces which the higher law has to contend against and overcome. In each order the education of the faculty is suited to the mode and conditions of the future life. In the infant, the lower nature should be so treated as not to beget in him the faults which are virtues in the lower order of being. The agency by which this treatment is administered is that of *tutors*. Their conduct and example are the appointed means by which the ideas of moral principle are first infused into the receptive mind of the infant.

It follows, therefore, that every human being derives his first ideas of moral principles, and his preparation for the future prevalence of his higher nature, from the feelings enkindled, and the thoughts suggested, by the treatment he receives and the conduct he witnesses on the part of his tutors. These are the means of diverting his Thoughts from the otherwise predominant feelings and appetites of his lower nature. On his tutors, therefore, and not on any inherited ideas, rests the heavy responsibility of the issues of their relations with him. If they would learn the result of neglecting these duties, that is, of leaving the infant to the unrestrained impulses of his lower nature, they may behold it in the lower animals, where it is exhibited in its own proper colours, without any of those cloaks which human law, social obligations, and fear of consequences, throw over its manifestation in the human life. There, in the brute creation, and also, alas! in the dens of iniquity and haunts of crime, of our own Christian country, are exhibited the natural workings of the lower nature where it predominates. In these latter scenes may be witnessed the neglected or ill-performed work of tutors in its *open* manifestation; but there are also *hidden* results which, if less flagrant, are not less terrible, known only to Him who sees the inner pollution of suppressed Thought.

In the judicious training of the infant Will, in its earliest manifestations, lies the germ of that true, spontaneous obedience, elsewhere discussed, which forms so important a feature, first in the relations between teacher and pupil in the later stages of tutelage, and afterwards in the conduct of life. Faith in tutors, the first step towards faith in God, is also first inspired in the very earliest stages of baby helplessness, by such treatment and consideration of the susceptibilities of infant nature as would attract and win it. If these important modifications of the lower nature be not effected during the plastic and dependent period of babyhood, every step in the progress towards independent action will increase the difficulty of planting them deep and fixing them potentially into the after-life. Hence the conditions of babyhood are providentially ordained and are the best that could be conceived for that end.

The practical effect of the treatment of children in their earliest infancy is discussed in a succeeding chapter. At present we have only to note, in a preliminary way, the fact, that the functions of the thinking faculty are called into exercise by the conditions proper to each of the periods of life, as they succeed each other; and that the nature of the product, Thought, will necessarily be determined by that of the subject-matter on which they are employed. It follows, therefore, that, whilst corporeal feelings and susceptibilities are predominant, the higher order of Thought can only be suggested by the treatment of tutors, and, if not so suggested, it will be wanting.

With the close of the period of babyhood the despotic reign of sensibilities and appetites terminates. A fresh dynasty assumes superior rule over the realm previously established. The subjects, Thoughts, remain unaltered in their character, but they own a new master. The state may be lower or higher in its nature, good or bad, loyal or rebellious, but whatever it be, it comes under a new power. The faculties of body and mind, and the organs of sense, have been growing stronger gradually, and now demand employment. Their new ruler is *activity*—constant, untiring, restless activity. Limbs and senses are exerted in every possible kind of tentative effort. During this stage, which we distinguish from babyhood, the one preceding, by the term *infancy*; the faculties, both corporeal and mental, make great progress. Tentative efforts are rewarded, sooner or later, with surprising successes, and still more instructive failures. From all alike new experiences are acquired, and new ideas formed. Every new accomplishment affords a fund of joy to its possessor, though not always so to tutors. The functions of the thinking faculty are now so much engaged in directing activities of all kinds, and with the novel sensations and ideas those efforts produce, that the deposed despot has a diminishing influence in the state, notwithstanding vigorous assaults on his successor, and some temporary defeats. It is during this period that oral language is acquired, but the progress made is mainly effected by means of imitative impulse, and increased power and range of observation and discernment. Language is as yet but little used except in the way of imitation, and is but very partially understood till near the end of this term.

With the free use of elementary language a new rule is established; not that the old state, formed under previous powers, loses any of its subjects, but only that they all fall under a new sway, that of *inquisitiveness*. During this period knowledge, experience and the resources they develop, all multiply and suggest aspirations for *independence*, which is the predominant force that exercises sway during school days, the term which actually precedes the state of independence assumed when the youth launches out on life's career, freed from the direct control of tutors and governors.

The course of the development of the faculties, as

shown in the succeeding chapters on their natural history, exhibits the way in which the functions of Thought are brought into use in the early stages of life. Such are the conditions under which the thinking faculty is trained and matured. The form and inclination received during the plastic period of youth, from babyhood upwards, will depend on the subject-matter on which the functions of Thought have been exercised, whether this have been suggested by the lower nature and the haphazard influences of environment, or have been of the higher class, judiciously infused by the treatment and example of tutors. On this will depend whether the sensuous, mental or moral element prevail, and this will decide the relation of the youth towards external persons and things, as embodied in temper and disposition. With such a capital of formulated Thought, and with such principles as he may have imbibed during the term of his tutelage, he will embark upon his career of independent action. Whatever may be the grade, occupation and mode of life to which he may have to devote his faculties, this will be his equipment for the great and decisive conflict between his lower nature, now sharpened by new desires, and the higher law of spiritual authority, which will decide his ultimate destiny.

Having sketched in the foregoing part of this chapter the functions of Thought, in the order in which they are called into operation, and exercised during the period of tutelage, whilst the faculty was in process of being formed and matured, its powers developed, and its direction determined, we must now consider the purposes for which this long course of preparation was destined, with reference to the future exercise of those functions in the independent and active career of the future business and duties of life. These purposes relate primarily to the maintenance of the social position, the supplying of the material wants, and generally to the exigencies of the lower nature of the individual man, conformably with the obligations imposed upon him by the moral principles of his higher nature, in relation both to himself and his neighbour.

Beside these purposes, there is also a higher and less obvious one disclosed in man's position in, and relation to, the order of external Nature.

All these purposes are inseparably connected and associated together in man's career in the present life, but will be more conveniently considered severally in relation to the functions of Thought, for they belong respectively to the higher and lower domains of his complex nature. The world of animated Nature subsisted for long ages under the domination of the lower law, anterior to the introduction of the higher, and it might have continued in that state for all time. Mental operations progressed during that period of the world's history to a considerable development, before the moral element was introduced, and the higher functions of Thought were called into exercise. Thought was long confined to purely material purposes in the brute races, and still continues to exercise its necessary offices in the lower nature, both in the lower animals themselves and in man. The operations of Thought in the higher domain do not diminish its functions in the lower conditions of material being, but they involve a new application of mental activities, and of the functions to which they are directed. This new purpose was to enable man to discover and comprehend something beyond the material aspects of Nature, and to penetrate the mystery therein revealed of the origin and authorship of the creation. As the material universe exhibits the works of the Creator, the higher nature was created in man that he might thereby comprehend the evidence those works contain, and discern in them the revelation of Himself.

The Lord God beheld the work of His hands, the heavens, the earth and all that therein is; and it was very good. All nature glorified Him; and He had pleasure in contemplating the beauteous creation, but it comprehended Him not nor loved. So He created in man a likeness to Himself, and consummated the material creation by establishing therein a connecting link with the spiritual world. By giving man a command, He placed him in direct relation with Himself, invested him with a higher nature than that of the material world, and endowed him with the capability of receiving a spiritual life. Hence man, bearing the likeness of the divine impressed on his human nature, was enabled to comprehend and appreciate the wisdom, power, and beneficence of the Creator, as these are manifested in the work of creation. Thus endowed, he alone of all creatures can offer to his Maker the spontaneous tribute of voluntary obedience, service, and love.

Hence man, by virtue of his higher nature, is raised to a position superior to all the conditions of the previously existing order of creation, and is capable of attaining to the honour and dignity of companionship with the Most High. This ultimate destiny, however, is to be reached through progressive stages of the development of his higher nature, from its infancy to its maturity in each individual, and also in the collective human race.

To man, endowed with the higher nature, is given the commission to subdue and exercise dominion over the earth and all the lower order of Nature therein. To him is confided the work of establishing the supremacy of the higher over the lower order prevailing in the world. In this earthly domain are the higher functions of Thought to be first exercised and developed. In this natural and material world he is to seek and to discover a Spiritual Head therein revealed. In the works of His hands and in the ways of His providence, in the harmonious laws, the beneficent purposes, and the wonderful adaptations of the material universe, man is to discern the Divine Author. Herein is the highest function of the thinking faculty.

Man's dominion over the external world is but a part of his commission, the rest, which is typified in the subjection of the outer material world is in the subordination of the lower nature in himself. This is the great purpose of the vicissitudes of life, its duties, labours, and issues. The lower nature, as it is exhibited in his environment, is the battle-ground on which the higher is to achieve its triumph over the lower nature in himself.

The bodily and mental activities, occupations, and duties of life are the opportunities, and the functions of Thought are the means, by which the great purpose of life, the subjection of the appetites and passions, is to be effected. Whatever may be the social position of the individual, whatever the nature of his occupations, or the circumstances and conditions of his life, the theatre of the world and the part he plays in it constitute the sphere in which his higher nature is to exercise its function, develop its character, and either win its victory or suffer its defeat. Every situation in life has its own special temptations, immunities, trials and opportunities; but in all alike the same appetites, passions and animal nature present the same field for the operations of the higher law. Diverse as are the circumstances in which the various classes of men acquire their experience, they stand on equal terms for the exercise of virtue; and those high principles which attest the worth of character are often most strongly developed in the ranks of the poor. Lazarus, with all the disadvantages of poverty, trial and affliction, and notwithstanding the mean circumstances of his sorrowful life and preoccupations, attained a far nobler position than his rich benefactor. In like manner all that contributes to true happiness is as much within the reach of the poor as of the rich. There is, in fact, as much real contentment, peace and thankfulness in the cottage as in the palace. Thought may be exercised upon circumstances of a widely different nature, but the practical issue does not correspond with external conditions. The enjoyment and the utilization of life are equally available for all, and depend far less on opportunity than on the use that is made of it. Character, which represents the work of the thinking faculty and the inner life, often attains its highest developments in the poorer classes. Our blessed Redeemer chose to assume a very humble position in this life. His companions were fishermen, His friends peasants, and His circumstances sorrowful.

Yet these were the conditions in which His exalted virtues and sublime character were exhibited to mankind.

The functions of Thought are exercised in all the positions, duties and occupations of life, whether concerning material interests or mental attainments, the trials or enjoyments, the inner self or the outward conduct, on what we are, or on what we seem to be. They determine the relation which, in the end, will subsist between the higher and the lower nature. Successively, through every phase and period of life, this is the ultimate purpose for which they were intended. They are the means whereby the individual self is formed and developed from the whole life, from infancy to the final dissolution of the material nature.

It would be inconsistent with all that is known of the works of Nature to suppose that a condition attained by gradual development through a life-long labour should suddenly terminate when it had reached its highest point. That would imply that the process of education, carried on through the varying vicissitudes and phases of life, was, so far as regards the individual himself, a purposeless waste of energy. Hence natural religion, even without the aid of revelation, would suggest, from the evidences of external nature, that a future state of being must have been contemplated for man by his Maker when He consummated the material creation by endowing him with a moral nature, and thus establishing in him a spiritual relation with Himself.

Seeing that, whilst the higher element of man's complex nature is united with, and partly dependent upon the lower, it is engaged, throughout the whole term of their union, in the work of subjecting that perishable element and obtaining its own supremacy and independence, it is both reasonable and natural that the higher nature, so freed and disencumbered, should rise to the enjoyment of the fruits of its lifelong struggle, to a new state of existence in a spiritual world, and to a further progress in a purely spiritual life.

The history of the natural life, from its dawn in infancy through all the phases of its transient and temporary experiences, up to the maturity of its mental and moral attainments, including the conflict between the higher and lower elements, is all indelibly photographed and immortalized in the collective body of individual Thought. Hence, when the material body is resolved into its component elements and dissipated, the body of Thought will rise a spiritual body, in which the natural life will be inscribed in imperishable form.

In like manner, whilst the material universe passes through the temporary phases of its development, from chaos up to its final and inevitable dissolution, when it shall have fulfilled its appointed cycle of change, its climax will be attained in the collective Thought of the human family, and it will survive in that spiritual form when the material body shall have been resolved into its component elements.

As in the individual man the corruptible nature dissolves, and rises again incorruptible in a spiritual body of Thought, so also the world of material nature, when its material body shall have fulfilled its destiny and reached its dissolution, shall also rise again spiritually embodied in the collective Thought of man, in which all its phases, forces, and evolutions, the sum total of its natural history, will be embalmed, and its development shall attain its climax, imperishable and eternal.

The function of Thought in the mental work of the human race, and the part it has fulfilled in the progress of the world's development, may be well observed by comparing ancient with modern phases of knowledge, or, in other words, the fanciful conceptions of loose conjecture with the facts of Nature and the inferences and demonstrations thence derived. Unfortunately for the bearing of mental work on religion, the Churches of all ages have too generally and too long overlooked the fact, that particular truth can never be divorced from the general body of truth. They have resented the intrusion of truth from without the pale of church doctrine, to their own loss and disadvantage. In proscribing Galileo and compelling him to recant, the Church of old did not prevail over the forces of Nature but only deferred for a time the irresistible admission of demonstrated truth. Science, so repelled, has pursued its researches independently, and has gathered in a rich harvest of truth, of which the religious world has taken note reluctantly, only when it could no longer be resisted. Nevertheless, the best service that the researches of science have rendered has redounded to the advantage of religion. Supplementing the imperfections inherent in human language, and correcting such error as the infirmities incident to human agency entailed, science has interpreted and explained scripture by its own natural interpreter, namely, the *works* of Him whose word it reveals in the imperfect vehicle of human language. The *works* are the direct product of His own Almighty power; the *words* are the indirect utterance of Divine truth through the human tongue. This utterance therefore is amplified and explained by the higher manifestation of the same Divine authority. God is manifest in His works, revealed, not obscured, and it is the function of Thought to apprehend His nature therein. Man's duty is rather to interpret the word by the work, than to exalt the imperfect vehicle above the direct revelation of the Almighty hand.

The foregoing considerations show that the functions of Thought build up the individual self in man, the embodiment of his natural life, and that they fulfil the same office for the material world for each successive age, and will continue so to do until the material world and life therein shall have served their purpose and passed away. Hence it appears that the spiritual element in the world is built upon the natural fabric, just as the same element in man's nature is engrafted upon his natural life. In both the individual and the race the spiritual rises out of the natural, is born there, and grows through successive stages until the material nature is dissolved, and the spiritual returns to God who gave it.

CHAPTER IV.

ON LANGUAGE.

LANGUAGE consists of associated sounds or signs with ideas of objects and their relations and qualities and of actions done or suffered, &c. When these or any of them are represented by sounds they constitute oral or spoken language; and when by signs, such as letters, hieroglyphs, or other graphic symbols, written language is formed.

Oral language is composed of words, which consist of complex sounds produced by articulation. These articulate sounds are for convenience analysed into their component simple sounds, which are represented by letters of the alphabet, the chief use of which is in the delineation of the words in graphic form for the purposes of written language.

Language attains its full development into the forms so recognized in the human race; but the rudiments of language exist and serve important ends in the lower animals. In the form of cries of alarm, indications of the emotions of fear, affection, anger and the like, and calls, in the form of amatory invitations or as warnings of approaching danger, this rudimentary form of language is in more or less general use throughout the economy of animal life. Important as such use is to the lower animals, it is very restricted. Happily so for man, who would have had an insuperable difficulty in dispossessing the brute creation of their tenure of the earth, if they had had so powerful a faculty, and such intercourse as it would have afforded them.

The origin of language proper may probably be traced to the rudimentary forms in which it exists in the lower animals. Many words are, in fact, derived therefrom, as for instance, *coo*, from the cooing of doves, *grunt*, *bark*, *purr*, *whine*, and the like. Moreover, the sounds uttered by animals derive much of their significance from the tones and inflexions by which the simple sounds are made more expressive. Such suggestive utterances afford a sufficient basis on which the ingenuity of man, even in his savage state, might construct, by slow degrees, a language worthy of the name. And such an origin is a pleasing reflection to a worshipper of God, and a believer in His providential government of the world, because it adds to the ever-increasing number of evidences of continuity and unity of design in creation.

Whatever the origin may have been, however, it is certain that language was formed slowly and very gradually from very small beginnings, the zero being undoubtedly, as already suggested, very low in the scale of animated beings, and the progress never such as to entitle any of its rudimentary forms to the designation now attached to the term, until it received that development in man.

Language is acquired by the human infant by observation, inference, and imitation. His first utterances are of the rudimentary form already mentioned as appropriate to the lower order of beings; but his acquisition of language proper comes slowly. The process is not very difficult to perceive and to follow, but it is nevertheless of great interest as an important element in infant history, and one that deserves particular attention in these pages. Several important inferences may be drawn, and some facts are disclosed, by an attentive study of the process which are not otherwise obvious. The most important of these refers to the fact that children of all capacities contrive to acquire the use of spoken language at a very early age, notwithstanding the many difficulties involved in the work. These will be specified hereafter, but in the meantime it may be mentioned that the work is done by observation and inference, without any teaching, and also without any very perceptible effort. It may be thence inferred that learning is not naturally distasteful, seeing that it is in this matter voluntarily undertaken and always pursued successfully with no other stimulus than that of ordinary desire. Moreover, seeing how vital a factor in the process memory is, it may be inferred that this important mental qualification is never wanting in infancy. If, therefore, it should seem to fail afterwards, the fault cannot be laid on natural incapacity. Nor if the natural appetite for knowledge should cease, and if learning should become irksome, must the cause of the change be sought in the original condition of the mind, but in subsequent mismanagement.

Recollection does not go back far enough to admit of any reference to personal experience in regard to the difficulties encountered in the first acquisition of language. Hence, the perplexities and pains involved in the first efforts in acquiring the elementary knowledge attained in infancy are too little considered in the treatment of young children. Failing appeal to personal experience, the only resource is by means of a careful analysis of the necessary factors involved in the processes of apprehending, associating, remembering, and articulating. These prove that the work to be done is complex as well as strange. By-the-bye, novelty is itself an embarrassing element, which is too generally overlooked in considering and allowing for the efforts of young children. And this remissness is the less excusable, because adults themselves are continually reminded by their own experience how difficult it is to catch and clearly apprehend a quite strange word. Yet this obstacle is one which stands in the path of almost all the early undertakings of the infant progress, though it is as unperceived by tutors generally, as was the apparition that barred the way and excited the impatience of Balaam of old.

In the acquisition of language, the two processes of understanding and expressing an idea must be considered separately, for the former precedes the latter by a period of time, always considerable, and sometimes very prolonged. It must also be borne in mind, throughout the discussion of these processes, that imitative attempts to articulate words merely as *sounds*, without any idea at all of their signification, like the talking of parrots, is going on quite independently of language, in the proper sense, during a great part of infancy. In order to understand the meaning of any utterance, however simple, there must be on the part of the infant: (1) a clear apprehension of the sound or utterance; (2) a definite recognition of intention or purpose of the speaker; (3) a perception of the idea or subject-matter expressed by the utterance; (4) association of the utterance with that idea; (5) recollection of each of the elements so associated and of the association itself. If any one of these essential parts of the process fail wholly or in part it must be made good before the result, a proper understanding of the utterance and its intention, can be effectively attained.

Now, we must ask some indulgent consideration, and exercise of patience, in regard to this process, not for its own sake only, important as it would then be, but chiefly with reference to the workings of the infant mind generally, of which this process is a fair exponent. In all the intercourse between tutors and infants it is of paramount importance that the former should fully apprehend the standpoint of the latter, and consider what is actually going on in his unfurnished mind. The supreme difficulty in the art of teaching, is in perceiving, entering into, and appreciating, the real condition of the pupil's mind, and the limited and imperfect stock of furniture it contains. Tutors can hardly help assuming that matters which have become perfectly, habitually and mechanically familiar to themselves can be so absolutely unknown and utterly absent from their pupils. They have no recollection of having themselves had to acquire them, or of the difficulties of that distant and obscure part of their own personal history, and are, therefore, apt to regard such elementary acquisitions as a natural and necessary possession. Hence their impatience of the seeming stupidity, inattention, or perverseness which a perturbed, vacuous mind is apt to simulate. And, of all the scatterbrain influences that can close up the narrow avenues to a child's intellect, the most effectual is the exhibition of the teacher's impatience! The recollections of a highly sensitive person who in early childhood was too sickly and weak to receive the instruction usually acquired in that period, and who had, therefore, received it at a later period of life, of which he retained ineffaceable impressions, depict in pitiable terms the tortures he endured from the impatience too practically manifested by his teachers, whilst he was all the time vainly struggling with all his might to squeeze out of his poor store things which were not there, nor ever had been! A struggle with the impossible, a dead pull, breaks the heart of even a colt, and it goes harder still with the more impressionable sensibilities of a human subject.

(1) Resuming the consideration of the process involved in the understanding of an utterance, the *first*-mentioned factor, though a simple sensation in itself, needs some notice, for the senses in infancy are not very proficient, and need to be frequently appealed to before a fresh sensation is fully apprehended. So simple a word as *shoe* might, and probably would, require to be very often repeated before it would be definitely formulated in such manner as to be readily recognized and obedient to the recall of memory. How much more would complex words tax the power of apprehension! Hence there exists a long series of similar efforts before a sentence of several words would be grasped, as sounds only, without reference to their separate or joint meanings! The difficulty of grasping strange words is shown by the many repetitions of a foreign word which are required before it is apprehended even by an adult.

(2) Of the numerous sounds that are ever impinging on the ear from all sides, any one that

represented an idea to be communicated would require to be addressed to the infant, and to be consciously received as an appeal to his understanding. It must be received and regarded as having a special intent and relation to himself, in order to enlist his attention otherwise than as the ticking of the clock, the chirp of the cricket, or any other like sound of a casual nature.

(3) The subject-matter must exist in the infant mind as a formulated idea representing the same. The utterance and the subject-matter, being the principal elements to be connected, must be perfectly present in the mind before any connection could be effectively established.

(4) There being no *necessary* connection between the two, the act of associating them together is distinct from each, and yet is *the* essential element of the process. It must be complete and intimate, for, if broken, there is an end of any useful result. Any sound or word may stand for any idea; the connection exists only in the mind. All the significance of the word consists in its fixed connection or association with a definite idea. How often does even the *adult* mind fail to recall the name of a person or thing, though both are tolerably well known! How much more effort must be required in the infant mind, where so few ideas or their connections can have become familiar!

(5) Having the ideas of the subject-matter and of the sound or word representing it both firmly established in the mind, and their association together in the relation of principal and representative, these elements must be *remembered*, each and all; word or utterance, subject-matter, and the mutual relation of the two, must be obedient to the summons

of memory, or they are not available for the purpose of being understood.

The foregoing refer to the utterance of single words; but when several are used together in a sentence, however short, the process as described above is necessary to the understanding of each word severally, and their relation to each other has to be gleaned by the experience derivable from the circumstances of their use. The meanings of words, and their relations as parts of speech, cannot possibly be acquired in any other way. A very comprehensive knowledge of language and its use is necessary before a dictionary is of any service. Hence very frequent repetitions of words, separately and in their commonest relations, need to be heard, understood, and reflected upon before the simplest formulæ of language can be comprehended.

Now, the work involved in the processes as described in the foregoing paragraphs has to be done, and *is* done, by every child who has learned to understand the simplest sentence expressed in words. The only help the learner has is in his acute discernment of gesture, facial expression, and the circumstances of the utterance. No wonder, then, that a knowledge of language comes so slowly. The fact that all children except deaf-mutes make the necessary effort to acquire language seems to be due to natural desire to know what is going on about them. Even the domestic animals thus come to learn the meanings of some words and simple phrases.

All these factors are necessary to the understanding of any idea, equally in the adult as in the infant mind; but, in the former, both the elements and the process have become so familiar by constant practice that the whole act of comprehension is effected mechanically, without any conscious effort, excepting only in cases where one or other of the elements is strange, or by lapse of memory incomplete. The frequent occurrence of difficulty to the adult mind in such cases should suggest consideration and allowance on the part of tutors for the infirmities of the infant faculties. In the relations between teacher and pupil, even in the more advanced periods of youth, a reference to the nature of the process involved would often assist the former in discovering the missing link that obscures the understanding of his pupil, and save futile, misdirected effort to both.

It is needless to say that the process required for the understanding of any utterance is equally necessary for the expressing of it. In addition, several other factors are necessary.

Gesture, which is the first kind of expression to be understood before any knowledge of language has been acquired, is also the first to be used. All the earliest efforts of speech are plentifully supplemented by the aid of this powerful resource. Two or three words will often do duty for all purposes, without, of course, any proper significance beyond what they derive from accompanying tone and gesture.

In one instance of our experience one word only was made to stand for everything the child wanted. It formed the total stock of her vocabulary for a year or more. No one knew how or where she acquired it, but she spoke it glibly, for it was constantly in use. "Cowdadda," with certain grunts and gesticulations, was her only articulated utterance, and was her universal noun. Yet, notwithstanding her extremely limited power of *expressing* herself in language, she proved by her responsive actions that she *understood* a very considerable number and variety of words and their relations. She was a very intelligent child. Such evidence shows plainly that the power of understanding speech is not to be measured by the power of speaking intelligibly.

The additional requisites for expressing ideas in speech, after understanding of it has been acquired, are—(1) The wish or impulse to express some idea; (2) The power of articulating the representative words, already supposed to be well known; and (3) The marshalling of the requisite words, that is, the method of constructing sentences.

Respecting the *first* of these, the only interest attaching to it as a factor is suggested by the difference there exists amongst children, some being habitually reticent and others loquacious. As there is a natural impulse to employ a newly acquired power, a reticent disposition should be traced to its cause, and if it should appear to arise from listlessness (that fatal tendency, to which special reference is made in chapter xi.) it should be counteracted by suitable stimulus.

(2) Articulation often requires long and severe effort. Some persons articulate imperfectly all through life, and some children make prodigious efforts with very slow progress. In many cases the diffidence due to conscious imperfection in this respect retards the free use of speech by a considerable period. It is remarkable that with the strong impulses and powers of imitation in the lower animals so very few seem capable of imitating articulate speech, though so many of them are capable of understanding words and short phrases.

(3) The principle of constructing sentences—that is, of arranging the order of the words therein—differs so much in different languages that it is evidently an arbitrary matter, not depending on any necessary sequence, but on habit. It is therefore acquired by imitation and experience, and consequently by a slow process of observation and reflection. Practice makes it easy to most people, quite mechanical to some. Still, adult teachers, to whom it has become so natural by long and constant practice, should not fail to make allowance for their unpractised pupils. We plead the cause of children, and do not disguise our intense sympathy for the struggles of the juvenile mind in its efforts to obviate the want of proper systematic training, based upon a due appreciation of their difficulties and the magnitude of their task.

Hitherto we have considered language only in its oral or spoken form, which was in use in a highly developed shape for many ages before any graphic form of expressing ideas assumed the character of a written language. Originally suggested by personal and domestic wants and purposes in primitive times, oral language was gradually extended to tribal and more general use. In this form priests and minstrels handed down from generation to generation the traditions of each. Such was the original and very unreliable means of disseminating the ever accumulating store of history, fabulous legends, and superstitions which then constituted the only knowledge extant other than that of the immediate time and present experience.

The origin of the graphic forms of language is obscure. It was probably suggested first by footprints and such-like indications, and afterwards developed through successive rude forms into the more regular shape of hieroglyphs, first pictorial and afterwards symbolical. The analysis of articulate words into their component elements, and the representing of these by symbols or letters in the regular form of written language, are comparatively modern, and mark a great epoch in the progress of the human race.

The development of language in the individual follows an analogous course to that of its general history. Beginning in the infant with the simplest sounds and gestures indicative of the most elementary ideas, it is slowly acquired through successive stages of use and development from the spoken to the written form, and thus becomes the most influential agency for the further progress of the individual, as it is also of the race.

As the understanding of speech precedes speaking, so reading precedes writing. The learning of the letters of the alphabet is often a slow proceeding, notwithstanding many means are employed to familiarize them individually to the infant eye. Still, to children generally they are strange, crooked, unmeaning things. Learning them is a dry task, and during the whole of the tedious process very few children have the least idea what it all means. When taking the first lesson in reading, and the child repeats the letters *d o g*, he is told to say *dog*, but he sees no reason why they should *spell* dog, nor why they might not just as well stand for cat. In fact, it is quite common for the teacher who asks, " Now what do d o g spell ?" to be answered, " *Cat*," and this perhaps after having been told what they spelt. He does not see the connection between the letters and the word. It is of course only too obvious to the teacher, wherefore it seems to him that the child is heedless, stupid, or perverse. Pity it is that he cannot remember his own experience. It would be far easier and pleasanter for him to explain, even if it took him some time, what is the use and value of letters as component parts of words, instead of hammering ineffectually to drive the nail in head first. If a child be made to understand the use of letters and *the sounds they actually stand for, as well as the names they are known by*, spelling would become a sensible, if not an agreeable pursuit, instead of being a sheer and seemingly senseless drudgery. The lesson would enter his understanding naturally, instead of being thrust in by main force. Lessons are not *tasks* in their own nature, but may be made odious by injudicious teaching. Under ordinary circumstances, children never understand the rationale of letters and spelling until the knowledge has become superfluous.

The alphabet is not well contrived, and is needlessly difficult. The names of the letters do not accord with their literal values, and are not even regular. It was evidently not contrived for children of the age of learners of this day. It might well be reformed upon some more intelligible principle. Not one child in a hundred comprehends the use of it till years after its use has been established by practice. Why should the vowels be scattered here and there, all through it, in situations without significance? Seeing that the consonants have each a vowel attached, why should not the same vowel be used for all? And why should the vowel be placed sometimes before and sometimes after the consonant?

As *e* is the prevailing vowel sound, it might be applied to all excepting *s*, thus: *be, ce, de, fe, ge* (hard), *he, je, ke, le, me, ne, pe, que, re, es, te, ve, we, xe, ye, ze*. Q should not be separated from the *u*, without which it is never used in English. As the alphabet now stands, it serves its purpose, truly, but it involves trouble and time far more than would be necessary if it were systematically arranged with reference to the capacities of those who have to learn it. In the meantime, teachers would do well to consider the task it imposes, not in learning so much as in *spelling*.

Much of the difference in the rate of progress in children, which is usually referred to mental capacity, is due to habit and other causes. For instance, one who has been accustomed to note visible objects will learn a lesson from a book, table, or diagram with greater facility than one addressed to the ear. One organ, be it the eye or the ear, is in some children much more easily and effectually attracted than the other. Such peculiarities, of which the subjects are not themselves conscious, may cause grief and retard progress if not perceived by tutors.

The use of language is the most influential agency in the progress and dissemination of knowledge; and in the advancement, moral and material, of the human race. It is impossible to conceive what would have been the condition of mankind without the power of speech; for even the lowest tribes of aborigines have had the advantage in attaining their present barbarous position. It is the currency in which the commerce of Thought is carried on and the products of the thinking faculty are interchanged, multiplied, and accumulated. It is the means by which fleeting thoughts are embodied in permanent symbols, and disseminated from their sources to the ends of the earth. It is thus connected with the spiritual element in man as being the vehicle by which spiritual influence is exerted on material nature.

The several methods by which language serves the use of mankind are—

1. As a medium of intercourse by speech;
2. Mentally, in thought and memory; and
3. For written record and intercourse—that is, for presenting to the eye in a permanent graphic form the fugitive sounds of oral language.

Of these we advert only to the second, as the others would lead us too far from our purpose.

It is doubtful to what extent ideas are formulated and exist in the mind in the shape of words—to what extent, that is, men think in words. Some persons even go so far as to opine that all Thought originally reaches the mind in that form. This, however, appears to us to be a quite untenable theory, for there can be no doubt of the existence of Thought in the lower animals,* and it is equally certain that all sane children have a copious supply of Thought before they have any knowledge of language at all adequate for thinking in words. The difficulty many people have in expressing their thoughts, and which even the most fluent speakers sometimes feel in putting into words their fixed ideas, is a further proof of at least a very considerable independence between ideas and their verbal formulæ. We do not here allude to lapses of memory. These will be discussed in the chapter devoted to that branch of our subject. We

---

* See Notes 8, &c., in the Appendix.

now refer to thoughts present to the mind, but which the possessor finds it very hard, and sometimes impossible, without much circumlocution, to put into words. This is a fact of daily occurrence. No man needs to tax his *memory* in order to define what he means by a *gentleman*, for instance; but few men would put their idea into words without considerable Thought and some time. The same might be said of many complex ideas, and of all that large and influential class of ideas mentioned in chapter ii. pp. 28, 29, as indeterminate ideas.

It follows, we think, that there exists in all men's minds a considerable proportion of ideas that do not lodge there in any form of words or images. The fact is so universally recognized, that a man in the witness-box, or in any other situation in life, is freely allowed time to formulate his ideas into words. But the same indulgence is not always shown to children! They are expected to put their thoughts into words on the shortest notice, and their embarrassed expression of countenance is regarded as evidence against them. A further proof of the fact that thoughts do not all exist in verbal form is afforded by the common case of a quotation in which the idea is correctly expressed, but not in the same words as those in which it was acquired.

Nevertheless mental labour is in many ways abridged and facilitated by means of verbal expression; for instance, in such words as *dwelling*, which comprehends a great range of very different structures having a common purpose, or *falsehood*, which embraces so wide an area of signification in a single word. A volume might be written on the uses and advantages of language in mental opera tions, so great, so numerous, and so universal are they.

There are, however, inherent and inevitable imperfections in language, as a vehicle of Thought, arising from the manner in which it is acquired. As words derive their signification from use and experience, they do not convey to all minds quite the same meaning. This, having in each case been derived from personal experience, differs accordingly in all those cases wherein the object assumes different aspects. A very large proportion of differences of opinion and disputes of fact are due to the different meanings suggested by the different aspects in which the fact was seen. It follows that a considerable class of words exists of which the meanings are not apprehended alike by all who use them.

Again, as language originates with the wants and ideas of a particular time, it needs to be enlarged and modified to suit the different uses which spring up anew as knowledge increases and wants multiply. The coining of new words to meet new requirements is an every-day process. Hence, in reading books of a bygone time, the terms used therein need to be considered by the light of the period in which they were written.

In endeavouring to exhibit the successive phases of the rise and progress of Thought in the human mind, we have found it necessary to divide the term of youth from babyhood to maturity into four periods for convenience of description. These divisions are quite arbitrary, as there is no natural division available for the purpose. Age is inapplicable, because it does not mark the same progress in all cases. We have therefore adopted Language as the basis of our division, as being the best exponent of both the progress made and the power by which the further advance is to be effected. The *first* period, which we distinguish as that of *babyhood*, extends from birth to the first efforts to utter words as such. The *second, infancy,* comprises the very active period during which oral language is acquired and brought into use along with that of the limbs and organs of sense. The *third, childhood,* includes that inquisitive stage during which speech is freely and intelligently applied to its proper use, and whilst, in the meantime, the elements of written language are being acquired. The *fourth,* the school-days, is that period during which the *youth,* duly qualified for book-learning, is ready to receive systematic education. Already aspiring to a state of independence of Thought and action, he is being prepared for the pursuit of the avocations, duties, and career of his term of probation in the world.

CHAPTER V.

ON TEMPER.

THE old saying, that a man's character is never fully displayed until he is crossed, contains an important truth, as showing how influential a factor temper is even when, in after-life, the obligations of social intercourse keep it under the control of the mature understanding and Will. Temper, in so far as it is not restrained by moral and social obligations, represents the general attitude or relation of a man towards his fellows. It is the exponent of the relation of his mind to theirs. Being the product or effect of what he has himself experienced from others, it shows the habitual state of his feelings towards the world generally.

In adults outward manifestations of bad temper are generally restrained by the necessity for self-control. The actual mental condition of the individual is therefore to a great extent concealed from his fellows in the secrecy of his own breast. To the world at large a man's temper is of little moment, except in so far as it is displayed in his outer conduct; but to himself its vital importance consists in the struggle within which it continually imposes. This is far more influential on his mental condition and character than any occasional lapses into which he may be betrayed. It is therefore in this individual and personal aspect of the question, rather than as a social element, that we are chiefly concerned.

Hence, temper influences the habit of thinking, and affects the whole tenor of a man's mental condition, as well as his relations with the world at large. It imparts a direction to the current of his ideas and reflections, gives a colouring to the circumstances of life, a bias to his judgments, and, generally, a tone, for good or for evil, to his whole character. Being, as it is, so important an element of right thinking, feeling, and acting, a necessary ingredient in all conduct, it is highly desirable that the history of its origin and progress should be carefully studied as a preliminary to the history and training of the human mind.

The practical effect of temper in daily intercourse is manifest in the manner assumed towards those persons whose temper is known. To one of calm and amiable disposition, a frank confidence is almost involuntarily conceded ; but, on the other hand, one approaches a person of uncertain or irascible temper with precaution partaking somewhat of distrust. No one would knowingly prefer a request or submit a proposition to a man of ill-temper when he was in an irritated condition, because it would not be expected that either would be treated quite reasonably or upon its merits. The effect of temper is therefore a regular subject of consideration in intercourse between fellow-men. Though a state of rage is, in the very worst cases, only occasional, and not perpetual, yet the habitual state of mind and mode of Thought in a man of infirm temper is unfavourable to sound judgment and unbiassed opinion, because it imparts a colouring, a personal character, to his ideas. In the later periods of life, when moral principles have become predominant, or when the amenities of social intercourse prevail, the manifestations of temper are controlled, and may be partly suppressed; but in childhood, before such restraints operate so effectually, outbursts are more frequent and more troublesome. Temper is therefore more observable, and also, in a social point of view, more inconvenient, in this than in later periods of life.

Being in its nature a state of the mind, and produced through the feelings, temper affects, directly or indirectly, every idea that enters into or proceeds from it, and is manifested in forms which correspond to the nature of the feelings in which it originates. The particular bias may be, and actually is, very different in different cases, giving rise to every variety of result. Moreover, the same causes would not produce the same results in a phlegmatic as in a sanguine or bilious temperament or corporeal constitution. Resentment, for instance, may beget in one a morose, and in another a violent temper. Variety of corporeal susceptibility will give rise to variety in the result. Be the temperament or constitution what it may, however, sanguine, phlegmatic, or what not, it cannot operate upon itself; it is but the recipient of external influences. These will determine the nature, those the degree, of the influence of the particular feeling in each case. There is therefore amongst men every variety of temper, bad and good, sullen and lively, amiable and vindictive, docile and perverse, *ad infinitum*. Be the temper what it may, however, it is the natural result of natural feelings. There is no magic about it. It originates in, and is developed by, causes regarding which there is neither mystery nor doubt. This conclusion, based on evidence of observation and long experience, will be questioned by those who have accepted on trust the too popular idea that temper is inherited. In a later chapter the subject of inherited tendencies is discussed at length. In this place we allude only to its bearing on temper.

The susceptibilities of human nature are affected according to certain fixed principles, which operate invariably as regards the direction and character of the result, but with certain limitations as to its force. In other words, every man will be pleased or vexed, soothed or irritated, according to the treatment he receives, though one man may be more readily and deeply affected than another by the same treatment. The result will be the same in kind, though different in degree. There is no special qualification necessary, for instance, for the feeling of resentment. Any human being may be exasperated, though one may be more sensitive and another more imperturbable, one may possess more self-control or have more regard for consequences than another; but no man is beyond the reach of persistent and determined irritants. Ill-usage will assuredly provoke dislike, and, if it be continually repeated, it will as certainly produce settled aversion. Treatment necessarily produces a corresponding state of mind, whether of attraction or repulsion, affection or dislike, temporary or permanent. Every day's experience gives practical proof of this characteristic of human nature, and shows, further, that *ill*-treatment is more keenly felt, and more lasting in its effects, than indifference or kindness. Such is the universal effect of the conduct of men upon their fellows. Hence it follows that the state of mind of every individual towards his neighbours corresponds with their conduct towards him individually or collectively, or, rather, with his view of their conduct, modified by reason and experience.

In adults, regard for appearances and other considerations restrain or modify the outward manifestation of their feelings, and conceal them, to a greater or less degree, from observation. The actual and natural effects are, therefore, not so manifest in them nor so easily recognized. Even in children, feelings are not always outwardly manifested, for in the very early stages of life they have no means of expressing them. Their limited resources of expression, crying and struggling, have to answer all purposes, and do not serve to distinguish between their corporeal and their mental sufferings. In later periods of childhood, the repressive measures vigorously exercised by tutors tend to keep the inner emotions from outward manifestation. In the human subject, therefore, of all ages, the feelings naturally evoked, secretly entertained, and influentially present to the mind are not fully exposed to view in their appropriate form.

In the lower animals the outward manifestation of feeling suffers none of the restraints imposed, directly or indirectly, by man's higher nature. In them, therefore, are exhibited the *natural* operation and effects of treatment, divested of all those complications which obscure them in the human subject.

Much may be learned by observation of the lower animals, in consequence of their unrestrained manifestation of feeling in its proper colours. In them are plainly and undisguisedly exhibited the natural working and effects of treatment, not only as regards the origin of those feelings which constitute *temper*, but also in the various devices and expedients which go to make up the inestimable art of attracting and establishing confidence. The lower animals attain the use of their powers of action and expression much more quickly than the human subject. The influences of their early training are simpler, and therefore the effects on the young scion of the treatment of his parents are more readily and certainly traceable to their respective real causes. Hence the high value of the evidence they afford to the student of nature generally, and of human nature especially, in which the animal, sensuous nature is inseparably bound up with the higher, moral nature throughout this present life.

In the Appendix a number of illustrations are given of cases, adduced from our own actual observation and experience, in which, by means of treatment alone, where no other influence could have interfered to affect the result, animals in whose nature there existed no tendency to vice or ill-temper were made vicious and intractable, or sulky and stubborn. Other cases show that the effects produced were such as to transform both the natural and the habitual tempers of the animals in question.

Each case possesses some point of special character, and the whole collectively afford conclusive evidence of the efficacy of treatment alone to produce in the subjects all those conditions which are commonly understood by the term Temper.

The case quoted in Note 1 is that of a remarkably fine old charger, "Blackthorn," which, after a long term of useful service and excellent behaviour, became in a very short time transformed into the most vicious and unnaturally fierce beast we have ever heard of in horseflesh. A gentleman, hearing this animal described, declared that he could not believe such ferocity possible in any horse, and he paid very dearly for his incredulity. Now, the remarkable points in this case were, *first*, the very extraordinary change of the animal's character, and, *secondly*, that it was discriminating, for the change was only in regard to Europeans. To natives the horse behaved with his former docility. He was to the last period of our acquaintance with him a fine serviceable beast when on the road, but the rider, if a European, required to mount and dismount with a native groom at the horse's head. The whole demeanour of the animal testified to the fact that the change of his disposition, terrible as it was, arose and terminated with the abuse and torment inflicted upon him. He was to the last free and willing at his work, and docile with natives, but irreconcilable to persons associated by their complexion with the persecutions he had suffered.

A case in some respects analogous to the one just mentioned is that of a little girl, a lovely child, of a sweet temper, intelligent, and animated. Her tormentor was an inconsiderate youth, who carried on a game of play to a very unreasonable length, and did not perceive the effect his prolonged teasing was producing. The strain at last exceeding the power of endurance of the gentle child, she suddenly sprang upon her tormentor, seized him by his whiskers, and, uttering a shriek, sank into friendly arms in an outburst of tears. This incident was temporary in its first effect, but the child did not fully recover from its sequelæ for some years. Eventually her former temper prevailed; but, had such treatment been repeated from day to day persistently, there can be no reasonable doubt that her temper would have been as effectually and permanently ruined as was that of "Blackthorn."

Another instance in which a settled habit of good temper was changed by treatment is mentioned in Note 2 of the Appendix. "Juniper" was a really noble animal. His owner made long journeys with him, rode him hard, but treated him with uniform kindness. His beautiful temper may be judged by the fact that, on dismounting on one occasion, and throwing the rein upon the horse's neck, the rider left him at the door, and was no little surprised to find the animal was following him up the steps into the veranda. He would come into the dining-room by himself, walk round the table, taking bread from any one who tendered it, and then leave the room in an orderly way, and have a fling outside by way of frolic. His gentle nature, high spirit, and fine paces made him a very valuable and serviceable beast. A change coming over him, however, with unmistakable indications of petulance and ill-temper, his owner discovered, after much inquiry, that a new groom had been using the curry-comb, which so irritated "Juniper's" fine sensitive skin as to eventually excite his temper. The abolition of the offending treatment restored the former habit.

One other instance in which an affectionate nature and habit were entirely changed by ill-usage is that of a pet monkey, which acquired a most ferocious temper, contrary to its own nature, and entirely opposed to a long habit of behaviour.

Accustomed for years to the utmost familiarity with his owner and the family, he became by a short term of teasing so very savage that he had to be killed.

In the foregoing cases good-nature and amiable habit were changed by various kinds of abuse to unnatural ferocity and vice. Cases of an opposite nature, in which natural ferocity is subdued and a peaceful and affectionate nature is acquired, may be seen in the domestication of cats and other members of the feline tribe. Cats are amongst the commonest of pets, and it sometimes happens that their own proper nature is so completely changed that they lose their predatory habit entirely and will scarcely notice a mouse or rat.

In this connection a marked difference which deserves notice exists between those animals whose proper nature is subdued by force, and those in which that nature has been overcome by kind methods of treatment. Those in which the natural propensities are only repressed are always liable to seize an opportunity, when the vigilance of the keeper is relaxed, and to resume their real unchanged character; whereas those which have been attracted by kindness will only resume their natural ferocity on being severely provoked or abused. A curious instance of the necessity for continued repression to enforce discipline may be here mentioned. A horse whose utter obstinacy had been always overcome by severe chastisement was running in a coach. All the time he was going, the driver continued whipping him with a heavy thong. A clergyman passenger remonstrated against this seeming cruelty. The driver replied that if he discontinued the whipping the horse would stop. The clergyman insisted, the whip was put up, and the horse stopped dead. For an hour and a half we had to stay in a burning sun, trying by every means we could devise to get the beast persuaded, but in vain. Another horse had to be procured. Lion and tiger tamers are well aware of the necessity for unrelaxed vigilance in regard to animals that have been subdued by force, and in which, therefore, the real nature is only repressed, and is ready to re-assert itself whenever opportunity may serve.

The foregoing illustrations are supplied by our own personal observation and experience. The history of each is known in all essential details, in such manner that the very striking results produced could be clearly traced to their several causes without the possibility of any others interfering to complicate them or affect their character. The natural history of the lower animals would supply innumerable illustrations in confirmation of the facts exemplified; but these, of which all the particulars are fully known, will, we hope, be deemed sufficient to prove satisfactorily that treatment is capable of creating and establishing a temper corresponding thereto, even in cases where natural organization and propensities and settled habits of long standing have to be overcome and completely transformed.

In the treatment of their young, the lower animals, even those of ferocious nature, exercise great forbearance, and rarely exhibit anger. On occasions of such unwearied provocations as a kitten or puppy may persist in practising on his parent, he may elicit a snarl or a snap; but in the worst cases it goes no farther. We have often seen the patience of a brute parent withstand a persistence of playful abuse that would have exhausted that of any human parent. Hence bad temper is rarely developed in the lower animals by the treatment of parents. In all the instances above quoted in which ill-temper was induced, it was caused by the abuse and ill-treatment inflicted by man!

In these illustrations the *natural* effects of treatment are exhibited in their bearing on the lower animals, whose sensibilities are of a low order. Seeing, however, that the effects are in all cases produced through the medium of the feelings, and that they are, in fact, the influential factors in the result, it follows that the intensity of the temper produced will be more or less directly proportionate to the sensitiveness of the feelings operated upon.

Applying the experience afforded by our illustrations to the case of infant life, they show that natural propensities, deeply fixed in the constitution, may be effectually overcome by treatment, even where this necessarily operates through the medium of a low order of sensibility. How much more effectually, then, would treatment prevail in the more susceptible nature of the human infant!

The highly sensitive condition of the corporeal feelings of the new-born babe, his pressing wants and utter helplessness, must necessarily supply the first subject-matter of his earliest mental effort—the first material on which his incipient faculties are exercised. His first ideas and experiences are those of ease or discomfort, satisfaction or disappointment, pleasure or pain. Without any means of voluntary expression, his whole consciousness is first absorbed by his internal feelings, and his first mental impressions must be of a nature corresponding to his corporeal suggestions. Feelings are at first the only influence, and for some time continue to be the dominant power of his nature. Therefore the first influence, and for a time almost the only influence, to which he is subject is that wherein temper originates and by means of which it is formed. It follows that temper is the first element of character impressed on the incipient mind of the human subject; and confidence in tutors is all but collateral therewith. The fine susceptibilities and utter helplessness and dependence of the babe, the two most influential characteristics of his nature, are, by providential arrangement, the means specially adapted to afford to tutors the fitting opportunity for the exercise of kindness and sympathy, which are the proper agencies for producing affection and trust, and for avoiding irritation and resentment. The wisdom of this providential arrangement for the initiation of love and faith, the soul of Christian principles, is also conspicuously manifest in the fact that in the human infant the powers which would make him independent are so slowly developed. Tutors have therefore the best possible means afforded to them by the mutual relationship between the babe and themselves, and by the protracted duration of that relation. The most adverse element that could be devised for counteracting this wise and beneficent arrangement is the theory that the babe derives his temper by inheritance, and that it is therefore a constitutional malady. By this fatal theory the babe is deprived, to a greater or less degree, of the advantage, and the tutors of the opportunity, designed by Almighty Wisdom for the happiness of both.

During the early stages of babyhood, the waking hours are occupied chiefly with the cares and attentions his appetites and his helplessness combine to exact from tutors. During this time feelings are not merely predominant, but constitute nearly the sum total of the babe's experience. They are always being considered or neglected, soothed or irritated, satisfied or disappointed; and the state of mind thereby produced must necessarily correspond to whichsoever of those feelings is most frequently and strongly evoked. In the midst of the ferment, the infant Will comes in to take a part in the incipient process of laying the foundation for the future edifice of the individual character. This new factor also operates on the same material which engages the exclusive attention of the faculties, and its force will be enlisted this way or that according to the prevailing current. Albeit, the operation of feelings and Will are carried on in the infant breast *unseen*, and too often totally unobserved and unwatched for. Hence it often happens that by the time baby has acquired the power of expressing his feelings by some of the commoner forms of gesture, he has also acquired some temper and some power of Will to show it. If these be, as they often and very naturally are, rather resentful and unamiable, he is supposed to have inherited a larger than usual share of the legacy of our great primogenitor. It is convenient then to forget the severe and daily recurring trials to which he may have been subjected during the processes of washing, dressing, and other like forms of (to him) unaccountable assaults upon his tenderest sensibilities.

In this connection we advert to the case of the

amiable animal "Juniper," whose beautiful temper broke down completely under the influence of the curry-comb on his thin fine coat. Here is a conclusive proof of the effect of exactly such a process as that to which many sensitive children are daily subjected. The fact that "Juniper's" very marked change of temper was caused by the daily repeated irritation of his skin was proved by the gradual cessation of his ill-tempers after the abolition of the offending instrument of torture. There is probably no single cause so effectual as that of the toilet in producing ill-temper in children, and, as it is a necessary as well as a continually recurring trial, it deserves the very careful management of nurses. It is not *necessarily* disagreeable, even to very young babes, if done gently and with tact. We have seen it done habitually in a manner to be rather a source of amusement than of irritation. Indeed, to children of a few months old it was a time of enjoyment, a game of play. A business-like nurse who professes not to understand "that sort of nonsense" will carry the whole process through, it is true, with a sort of air that seems very masterly. Poor baby is in her hands like clay in the hand of the potter, utterly suppressed and overwhelmed; but his is not the quiescence of ease and satisfaction, nor is the suppressed inner feeling at all like the outward demeanour. He may be conveniently cowed, or helplessly suppressed, but these are not the feelings which constitute or initiate a fine temper, nor is a Will that is forcibly overpowered effectually subdued. It is but pent up, to be hereafter manifested in such form as circumstances may determine.

Common-sense and common experience, if duly

consulted, would prove that every experience of life is influential as an item of the total sum which makes up the individual self. No part of the personal experience, from first to last, is lost. Through every stage of life, from birth onwards, all that is done or suffered is influential in producing the final result. And those experiences which first occupy the virgin soil of the mind are the most effective, because they are those on which all future ones are formed, and from which all others derive their colour and character. We have already shown that the first impressions are necessarily produced by feelings, and that these are the chief factors in forming temper.

If the keen susceptibilities of children expose them to some special forms of suffering, they are also eminently useful in securing to the sufferers a degree of tender care and sympathy they would not otherwise enjoy. These suggest and inculcate such lessons of love and moral principle as could not in any other way be made so practically intelligible or be so deeply impressed. If impatient and inconsiderate treatment excite the feelings to produce ill-temper on the one hand, love, sympathy, and consideration will, on the other hand, beget an affectionate, amiable disposition. The babe in his second stage will yield the natural fruit of the treatment he received in the first. According as he has been treated considerately and *consistently*, or has been thoughtlessly, impatiently, and fitfully regarded, so will necessarily his feelings have been affected and his temper formed. If petted at one time and snubbed at another, now treated as having both feelings and intelligence and anon as though he had neither the one nor the other, his temper will partake of the same fitful nature. As is the sowing such will be the harvest.

It will be freely admitted that crosses abound in every one's experience through life; and it is probable, though it may not be so generally allowed, that they are, in one way or another, very evenly distributed amongst all sorts and conditions of men. In individual experience, however, they are much more frequent and vexatious in early life, whilst children are subject to tutors, than they are afterwards, when actions are more independent, interference less obnoxious, and experience more influential. The crosses borne in childhood, especially when the powers of mind and body are beginning to be freely exercised, are of constant recurrence, and, though seemingly trivial in themselves, are peculiarly trying to temper. Operating as they do on sensitive natures, their weight is not to be estimated by their apparent insignificance, but by the feelings they arouse in those who have to bear them. Moreover, the conscious inability of the victim to escape, or to resist, aggravates the sense of irritation. Baffled resistance is very vexatious.

Hence it follows that the periods of infancy and childhood, when natural impulses are subject to the frequent, almost continual, action of the curb, are necessarily most trying to the temper. For then the feelings are most susceptible and least under control just when they are subject to their severest crossing. If, then, the Will and temper be not watchfully and considerately treated in those stages of the mental history, and if they be then allowed to acquire force, future crossing will necessarily need to be more frequently inflicted, and more severely felt. The sooner and the more effectually obedience is taught, the less will crossing and its attendant evils bear upon the young mind. This fact is, however, but too little recognized by tutors, who, regarding temper as an inherited characteristic, do not see the need for the exercise of those precautions by which it might be regulated.

When children have acquired the use of speech and some independence of Thought and action, the administration of repressive and corrective measures becomes more irksome to tutors, on whom that duty properly devolves. It often happens, therefore, that ill-tempers and bad habits are passed over at home, and the difficult and delicate task of correcting them is relegated to the schoolmaster, whose proper duty is thus made more distasteful to his pupil and more irksome to himself. School-life, where so much important work ought to be done, and where so much depends on the mutual relations of pupil and master, should be made as happy and smooth as possible, and be divested by all practicable means of everything that could interfere with the progress of the work. But, when home influences have been neglected, the pupil and master are at once placed in the position mutually most adverse to the fulfilment of their respective duties.

Unhappily, it is no uncommon thing for parents to threaten school to their children in such manner as to give it the character of a place of correction, thus unwisely inspiring in their minds a dread of school, and a feeling of hostility to the authorities there. The schoolmaster has then a double duty to perform, and has to sow his good seed amongst weeds and disorder. This transfer of parental duty cannot fail to impair in children that confidence in their parents which is the true foundation of filial respect.

The natural history of the moral principles of pupils so relegated to school for correction would generally disclose an unsatisfactory state of home affairs, where careless, inconsistent want of discipline prevailed and neglect of those principles on which good habits and right ideas of duty are formed.

CHAPTER VI.

ON THE NATURAL HISTORY OF THOUGHT.

FIRST PERIOD : BABYHOOD.

THIS period, commencing at birth, extends to that stage of the mental progress when the babe first attempts to utter words imitatively, with some idea that they have a meaning. It therefore embraces that part of the mental history before the powerful influence of language comes into operation.

At birth the babe possesses a corporeal frame endowed with keen sensibilities, and mental faculties, but has no power over either until it has been acquired by much tentative effort. Hence, though he can do nothing, he can suffer much. Life and action commence with the involuntary functions of the lungs, heart, &c. The use of the limbs and organs of sense is slowly acquired under the stimulus of sensations which are addressed to the sensorium, and are also involuntary.

The Will is stimulated into action under the same involuntary influences. The first efforts are very feeble and tentative, but by slow degrees the corporeal and mental faculties become amenable to control.

The mental faculties are first excited by sensations of appetite, satisfaction, pain, disappointment, and like personal feelings, and are soon afterwards attracted by perceptions of external objects and influences, also of an involuntary nature.

Hence it appears that the mental powers are first exercised upon involuntary sensations: either from within, as appetite or want; or from without, in the form of impressions of external objects, and especially of the treatment experienced from tutors. Voluntary action begins with the limbs and organs of sense, and proceeds gradually to extend to the mental faculties in formulating ideas and in reflecting thereupon.

In the early stages of babyhood the dominant influence of personal sensibilities is far more productive of ideas than are the results of voluntary effort. For, whereas the feelings are all the while in full force, the limbs and organs of sense are unpractised and but partially under control. Hence ideas of personal feeling are very influential before other kinds of idea are acquired.

OF THE CORPOREAL FACULTIES.—The limbs in early babyhood are hardly ever quite at rest except during sleep. In their involuntary restlessness the action of the Will is evoked; and the movements, which are at first spasmodic, become tentative, and gradually acquire direction and force. Incessant trials resulting in occasional successes are thereby stimulated, and by means of much repetition establish eventual control. By the end of this period the babe will have learned to walk, and acquired free use of his limbs and organs of sense. The eye and ear, which are the organs chiefly instrumental in attracting the perceptive faculty, soon begin to divert attention from the more persistent and pressing obtrusiveness of the appetites and feelings, and before the end of the period will have become effective agencies. In early babyhood impressions on the sensorium require some time before their significance is fully apprehended, and considerable effort is necessary to grasp and formulate them. Indications of effort may be observed in a fixed, wondering gaze, followed by joyous emotion when success is achieved. This effort is exacted by the inexperienced use of the organs of sense, and also by the novelty of the objects themselves, both taxing the unpractised faculties. It may be here mentioned that strange sights and sounds, even when presented to the adult mind, require special effort of apprehension and formulation. In such cases the difficulty is observed by the expression of the hearer's countenance, and he receives prompt aid from the speaker. In infants the effort to grasp a sound or sight often causes a baffled expression, which, however, rarely elicits help, and is set down to incapacity.

As the environment in babyhood is very restricted, and the same objects are constantly submitted to tentative scrutiny, very definite ideas are eventually formed of such familiar objects as the features of tutors. These are recognized after a few weeks, and are greeted with responsive smiles and gestures, especially when they are associated with sounds of the voice which confirm the evidence of the eye.

OF THE WILL.—Everything absolutely necessary to the life of the new-born babe is either conferred by involuntary vital action or is supplied by tutors. The Will is probably first prompted by desire to modify and control the involuntary movements of the limbs. Effort, which is a manifestation of Will, is indicated by movements and gestures caused by pain or other emotion, and in attempts to control and to use the limbs and organs of sense. Like all the other faculties, the Will acquires its force by exercise, and, being first exerted during the predominance of feelings, the ideas these originate are the first to engage its active power, and they continue throughout babyhood to supply its chief stimulus.

In the brute races, where might is supreme, and where no moral restraints interfere with the permanent ascendancy of the lower nature, the Will is encouraged and cultivated by parents in their young. In the brute, strength of limb would be of little avail without corresponding force of Will; wherefore a strong uncurbed Will becomes a creature which prevails by might. Manifestly, however, if the Will in the human infant be allowed to acquire its force under the stimulus of the animal feelings, the lower nature in him, instead of being subdued, will become dominant.

The fine susceptibilities and helpless dependence of the human infant place him so completely in the power of his tutors, and afford them such ample means and opportunities of acquiring his love and trust, that they may thereby easily mould his Will into spontaneous submission if they take due advantage of these conditions whilst they are in force. The importance of this part of infant training cannot be in any other way so strongly appreciated or so effectively enforced upon the attention of tutors as it is by the prayer, divinely taught and daily uttered, for that grand consummation when God's Will shall be done upon earth as it is in heaven. The subordination of the Will by culture, not by suppression, is thus shown to be at the root of all virtue; whosoever, therefore, sincerely indulges that high aspiration will recognize the necessity for a right direction of the infant Will from its first indication, and whilst as yet it is susceptible of being moulded.

OF THE MENTAL FACULTIES.—Perceptions of personal sensations are almost coeval with life itself, and for a time engross the mind almost exclusively. Throughout the term of babyhood personal feelings predominate, and therefore give the first direction to the exercise of the mental faculties and supply the chief subject-matter for Thought.

Personal feelings originate from within, suggested by appetite, pain, satisfaction, and the like; and from without, produced by the treatment of tutors and others. The emotions and ideas resulting from the conduct of tutors will naturally correspond therewith. If this be kind, considerate, and consistent, it will inspire love and confidence; but if otherwise, it will excite distrust, irritation, or other natural responsive feeling. In all cases, suffering is more influential than its opposite, and probably to a greater degree in children than in adults.

In this period wants are many and pressing, and are all the more deeply felt owing to conscious helplessness. The manner of ministering to them by tutors is generally considerate, judicious, and tender, but is often nevertheless inconsistent, irritating, and painful. The continually recurring washings and dressings are a fruitful source of trial and discomfort. Maternal solicitude has suggested many mitigations of suffering by means of safety-pins and other contrivances; but, after all that ingenuity has accomplished in this way, there still remains in these necessary processes a great deal to irritate the sensitive skin and tissues of young infants, and to try their tempers.

Seeing how the long-established amiable temper of an old horse was changed by such means (see Appendix, Note 2), it must be evident that the keen sensibilities of babes cannot be proof against similar influences. Hence it is undoubtedly the fact that most bad tempers owe their origin to inconsiderate treatment in infancy. Though the individual acts and the particular forms of mistreatment are not remembered, their results endure in the form of indeterminate ideas, and may affect the character more or less influentially through life.

Happily, the sensibilities proper to this period are quite as amenable to good as to evil influences. As soon as the babe is able to recognize the tones and inflexions of voice, and to distinguish their significance—when he acquires ideas of the love that underlies the attentions he receives, and comprehends the meaning of the caresses and endearments lavished on him—these influences begin to operate. His susceptibilities suggest other feelings than those of mere satisfaction and ease, and he becomes animated with love and trust, which supply a salutary kind of subject-matter for the exercise of his mental powers, and for his natural propensity of imitation.

The ideas of a personal nature formulated during babyhood are necessarily such as the circumstances and experiences of that period suggest. Such being the subject-matter on which the mental faculties are exercised, such will be the product in formulated Thought.

External objects, being perceived by means of impressions on the sensorium, depend upon the proficiency acquired in the use of the organs of sense, and in the power of the perceptive faculty itself.

Hence the first ideas of external objects must be of uncertain character, and very weak. For this reason many repetitions and much tentative effort must be required before any clear and definite ideas are formulated. It is doubtful whether objects are recognized as external until memory and reflection aid the perception. As the infant's environment is very restricted, and the objects first submitted to his senses are constantly repeated, the infirmity of his faculties, and his inexperienced use of them, are thus compensated.

The eye and the ear are the principal avenues by which sensations of external objects reach the mind. These organs operate very differently. Objects seen generally dwell before the eye, so that they may be contemplated and considered attentively for a certain time. Sounds, on the other hand, generally pass quickly, and afford but brief opportunity for observation. Moreover, objects seen are usually of definite form, colour, and aspect, whereas sounds have less distinctness, differing only in tone and loudness. Again, objects are generally seen in fixed relation of place, whereas sounds may emanate from uncertain sources, and are not always associable with the things they stand for or proceed from. The image of an object is directly connected with the object itself, but a sound does not so clearly appertain to its source. The sensations received through these different avenues are therefore very differently formulated, and the difference of the evidence they afford has a material influence on mental work and history, especially in infancy.

Music is the simplest form of sound, as it needs only to be heard, whereas sounds which have significance, like words, need to be also interpreted. An image explains itself, but a sound has no meaning, except what it derives either from its association in a sequence of sounds, or with some associated significance arbitrarily attached to it. Hence the image of an external object is more easily formulated, and more definitely inscribed on the mental tablet, than a sound, which is fugitive and, except by association, unmeaning. Hence music owes much of its soothing power to the simplicity of its nature, and the slight effort it exacts for its full apprehension. It needs no interpretation.

The product of the formulative faculty, the work actually accomplished during the stage of babyhood by means of this faculty in regard to external persons and things, in the restricted environment in which the ideas were formed, proves that good use has been made of the faculties; but it is still more marked in the increased facility of using them than in the products yet acquired by their use. The babe has already collected a good many ideas, and has still more notably enlarged his powers of making further acquisitions.

Memory.—As only those ideas which are subject to recall can be reflected upon, memory is the basis of all reflective action. An object present to the sense may be contemplated, but past sensations must first be recalled before they can be subject to reflection.

As memory is so important a factor in the development of the mental faculties, it forms the subject of chapter xiii., where it is discussed at length. In this and the succeeding chapters of the natural history it is treated only in so far as it marks the progress of the infant mind.

Evidence adduced in the sequel will show that in babyhood considerable power of memory is exhibited by various acts of recognition and reflection.

Reflective action is manifest in all those complex perceptions into which the recollections of past sensations enter and mingle with others afterwards presented to the mind. This is effected by comparison, association, and inference. Hence it follows that simple ideas are few and elementary, and are soon succeeded by others of a complex nature. But whereas simple ideas are definite, complex ones are, in babyhood, chiefly of the indeterminate kind described in chapter ii. p. 28.

Images of external objects, with some associated qualities, may be more or less distinctly and definitely formed in the mind; but complex ideas, involving reflection, must in this stage of mental progress, before experience and the aid of language have begun to simplify the processes of Thought, be chiefly of an indeterminate character. The prick of a pin, for example, will give a very clear idea of the pain inflicted, but the idea suggested by the circumstances of its occurrence will be indeterminate, because associated with others imperfectly understood. Still, the sense of pain will be definite, and the connected idea not the less influential though indeterminate.

Excepting such ideas as colour, form, inarticulate sound, and such-like elementary perceptions of a simple nature, all others are more or less complicated by reflective action; for even those simple ideas soon become associated with ideas of distance, dimension, and other relations suggested by inference or comparison with past experiences.

The recognition of a difference of voice and feature between the parent and the nurse, of gesture indicative of caress or rebuke, and of the feelings so indicated, all of which are expressed by children's gestures in babyhood, prove—(1) that the features of those persons had been clearly formulated; (2) that the changes of feature had been perceived; (3) that both features and their changes were remembered definitely; (4) that the emotions indicated were understood; and (5) that they were associated together. All these mental processes must have been performed, as every one is involved in the resultant responsive actions of the infant.

The feelings of want and appetite are early associated in the baby-mind with the corresponding sense of satisfaction or disappointment, ease or discomfort, &c. Hence, pains and pleasures of all kinds are formulated in association with the circumstances of relief or suffering, satisfaction or denial, consideration or neglect. The fact of such association is proved by expressive gesture. A smile might possibly be a mere act of imitation, but the apprehension of its significance is attested by unmistakable evidence of the manner of its occurrence.

Sounds and inflexions of voice, themselves requiring considerable discernment and proficiency in the use of the ear, are unerringly interpreted at the age of a few months, and must therefore have been associated with their proper meanings. In this act of associating ideas with signs is originated the first step towards *language*. For, although the first intelligible signs in which meanings are suggested are mere inflexions of voice, accompanied by gesture, in which the eye and ear co-operate, and are not of the nature of verbal or regular language, the act of association is complete in the recognition of their significance, and this is the basis of language.

Inference and judgment, being based on evidence, are little exercised by infants. Concerned almost exclusively with Thought originating in or bearing on their feelings, their inferences are mainly respecting the causes of the feelings they experience, and these inferences are often erroneous. Nevertheless, there is evidence in the demeanour of very young infants to show that they possess the faculty of judgment in some measure, and that they use it. The unaccountable preferences and aversions shown in this period involve some form of inference respecting the subjects of their preference or dislike.

Comparison is in itself an elementary form of inference, and is probably the first step towards a regular appreciation of evidence. Association, also, when it takes the form of attaching meanings to signs and gestures, implies an elementary act of inference.

TEMPER is the exponent of the relation of an individual towards persons generally—the state of mind inspired by the experience of the treatment he has received. It will therefore correspond with that treatment.

In the helplessness of babyhood, when wants are keenly felt but cannot be expressed, and when they are, in fact, often misunderstood, the relations between the babe and his tutors are very strained. In other words, much feeling is evoked, and a corresponding temper or relation of the infant towards tutors (who represent in babyhood external influences generally) is thereby established. This feeling is manifested by struggles and other forms of resentment on the one hand, and by demonstrations of joy, contentment, and satisfaction on the other, according to the kind of treatment or mistreatment, care or neglect, respect for or disregard of feelings, accorded to the child. The whole experience, of whatever kind, is productive of corresponding feeling, and this is embodied in the relation or attitude of the passive subject to active influences. Hence temper and disposition originate, and are often very firmly established, during this trying period of mental development.

FAITH is the fundamental principle of true obedience, and is the chief factor in giving force and value to example.

It is first inspired by the office of tutors during the helplessness and consequent dependence of the babe. In this lower aspect it depends upon the care, forethought, and love of tutors in the discharge of their duties, and it corresponds in degree with the treatment they accord. It is the parent of that higher faith which is the basis of all true religion and moral obligation, and owes its origin to the same feeling, for the paternal care and love of the Almighty Father supply the first impulse and the dependent sustaining power of the faith His creatures repose in Him. First, that which is natural, and *then* that which is spiritual, is the rule in faith as in the universe.

Implicit, in the first instance, owing to the dependent condition of the human infant, it will be confirmed, strengthened, and established as a rule of conduct on the one hand, or shaken and destroyed on the other, by the experience of tutelage.

OBEDIENCE.—True obedience is spontaneous—a desire inspired by confidence and trust to please those who exercise authority. In this stage of infant life, when need is pressing and power wanting, obedience may easily be forced. The babe may readily be overwhelmed; but forced obedience is but the illusive semblance of the genuine principle. True obedience is not the suppression of the Will, but its own spontaneous submission. It must be attracted, it cannot be forced. Hence the efficacy of trust or faith in the authority to which it has to yield its voluntary adhesion.

The importance of obedience in the later periods of childhood can scarcely be overrated. Hence the value and efficacy of faith in tutors in this period; for if obedience be not inspired in this stage, before the Will has acquired strength by habit and unrestrained exercise, there is little hope of any such result afterwards. The only hope then is that reason and moral principles may be influential. But if this fundamental principle be wanting, there is little reason to expect other moral principles to predominate over the lower nature so fortified.

Hence, babyhood is the golden opportunity for laying the foundation of this and other influential factors in the formation of the future character, and for furnishing the armoury by which the higher nature is to subdue the lower in the lifelong conflict between good and evil.

MORAL OBLIGATION is the relation of the individual to authority, and involves duty to superiors, in whom authority is naturally vested.

The first ideas of this relation are inspired by example, and by the natural exercise of authority on the one part, and natural submission on the other. The conduct of tutors as guardians, being based upon, and regulated by, a knowledge of good and evil, is, or ought to be, a constant exemplification of the prevailing power of the former. The moral influence and self-abnegation of the human parent forms a strong contrast to the example and tuition of brute parents, to whom moral principle is unknown and inapplicable. In them the lower nature is ever predominant. But in man that nature exists only as a necessary subordinate to a higher. Hence, whilst the lower nature is predominant in the human offspring, the prevalence of the higher is initiated and maintained by parents, as proxies. This is the nurture and admonition which naturally devolves on them, and is the means of establishing in the infant those principles by which he is hereafter to act for himself.

The idea of moral principle, the distinction between good and evil, originates as the infant observes and imitates, and thus imbibes the ideas inspired by the conduct of his tutors. This is not necessarily by special acts or intention of theirs, but goes on naturally and continuously in their habitual exemplification of the virtues suggested by their moral nature. Thus is the moral element gradually infused into the infant mind, to be thereafter nurtured in close connection with, and relation to, the lower domain, so long as the corporeal frame to which that domain belongs shall survive.

In no part of the providential dispensations of this life are the wisdom and beneficence of Divine ordinances more conspicuous than in the mutual relations and conditions in which parents and children stand to each other with reference to moral principles during the long term of the dependence of the latter on the offices and guardianship of the former.

HABIT OF THOUGHT.—The force and effect of habit in the domain of Thought is separately discussed in chapter xi., but the subject is mentioned in these chapters only so far as it bears on the particular periods in question.

Naturally, during the dominating influence of feelings, all those ideas and currents of Thought which originate therein become more or less habitual.

Temper is a habit. It is the relation of the individual to others, not at a particular time or on particular occasions, but his *habitual* attitude towards them. As a habit, it acquires its special force, and becomes a very influential characteristic of the individual in his general conduct, and in colouring all his own proper ideas, opinions, and judgments. Temper and its ally, disposition, have already been shown to be early initiated, and often strongly developed habits during babyhood. They are, in fact, the most important of those initiated in this stage of the mental history. In after-periods reason begins to operate in modifying the effects of treatment, but in babyhood feelings have undisputed sway.

SUMMARY.—From the foregoing sketch it will be seen that during this period, under the dominating influence of corporeal sensibilities, the babe will have acquired a free use of his limbs and senses by dint of constant exercise of both, and that he will also have made considerable progress in the use of his mental faculties, as is indicated by the following acquirements, viz.:—

1. A fund of experience supplied by his strong

perceptions of personal feeling, excited by sensations from within and from without.

2. His Will, excited by impulses of want, appetite, and sensations generally, will have received its bias, in one way or another, to a very influential degree, according to the way it has been treated.

3. An intimate knowledge of the few external objects and persons continually under his notice in his very restricted environment.

4. His stock of formulated ideas, necessarily derived from the subject-matter on which his faculties have been exercised, will correspond therewith; and will consist mainly of those originated by the dominant influence of feelings, along with definite ideas of certain familiar persons and external objects.

5. His reflective powers will be represented by acts involving memory, association, comparison, inference, and some elementary forms of judgment. He will by inference have obtained some ideas of distance, dimension, and other primary qualities of things.

6. Ideas formed under the influence of feelings often and strongly excited by emotions of joy and grief, satisfaction and disappointment, will be mainly indeterminate, resulting in a certain relation or attitude towards persons generally. This relation is exhibited in temper and disposition, and towards particular persons in the form of faith and obedience.

His ideas of moral principle, that is, of relation to authority, are as yet very elementary, derived by inference (the weakest of his immature faculties) from the conduct of his tutors. But, though rudimentary and but very imperfect, these ideas, being the first to direct his thoughts towards his higher nature, will be very influential in the ultimate formation of his character.

The effective or actuating result of the experiences of babyhood may be almost summed up in the condition which the Will and the temper have acquired.

This catalogue of attainments affords conclusive proof that, during the short term of fifteen or eighteen months of babyhood, a very considerable amount of mental progress has been made, and a great deal of effective mental work must have been done in one way or another.

However unoccupied and unconscious he may have seemed to be during those long hours when his incipient faculties were silently at work; though he may then have been regarded as having no more Thought than an animated doll; though he may, in fact, have been treated as such—caressed when convenient, and anon put aside as if he were insensible of neglect; now fondled, as if susceptible of love and appreciative of attention, and then treated as a plaything insensible of treatment—yet, at the end of this short term of incongruous experiences, the long list of baby's accomplishments proves how effectively his efforts have prevailed over the difficulties of the situation.

CHAPTER VII.

ON THE NATURAL HISTORY OF THOUGHT.

SECOND PERIOD: INFANCY.

THIS period, following babyhood, is that in which spoken language is acquired. By the end of this stage of the infant-history speech will have become familiar and some preliminary ideas of the graphic form of language acquired. Free use of the limbs having been obtained during babyhood, the natural and prevailing impulse now is to employ them. Feelings, being still sensitive and in strong force, supply the chief stimulus to the use of the corporeal powers. The dominant characteristic of this period is *activity*.

The further construction of the individual self during this succeeding term of infancy will proceed upon the lines laid down in babyhood. What was then done will constitute the basis upon which the future work will progress. What was then written on the sensitive tablet of the incipient mind there remains inscribed. It may be modified, but cannot be effaced.

The infant, by virtue of his experience in babyhood, will have acquired certain relations with the external world. To him the world was represented by his tutors and the few circumstances of his nursery life. These relations with the world outside of himself, and a very limited stock of ideas of definite objects, form his equipment for the little independent action he now commences. He is no longer absolutely helpless. The power he feels of acting for himself is the new stimulus of his nature, and prompts unceasing action. His nature is less plastic than before, and has already assumed the rudiments of a shape which will endure to some extent through life—whether of beauty or deformity, for good or for evil, will presently be seen. Hitherto concealed in his own breast for want of the means of manifesting it, it will soon become evident in the direction of his various activities. In these will be seen the disposition and temper impressed upon his nature by the treatment he has received during the most plastic period of his life. His behaviour to his doll will probably bear a strong likeness to that he has himself experienced at the hands of his nurses, partly, perhaps, unknown to his parents. All he has hitherto known and felt is the result of his own individual observation and experience; and he will now show what sort of lessons he has been learning while he was supposed to be doing nothing. The lines that now occupy the once blank page were chiefly inscribed there during those seemingly unoccupied hours of babyhood when he was deemed unthinking and inert. The virgin soil of his mind will be found to have been sown with seeds which may spring up in perhaps unexpected forms. The future cultivation or neglect of that soil will proceed along with those growths, whether for good or for evil.

OF THE CORPOREAL FACULTIES.—By dint of constant effort and tentative exercise, the limbs have now become amenable to the Will, and have acquired some strength.

The organs of sense fulfil their respective offices without effort, and are capable of a considerable amount of discrimination.

The corporeal faculties are now exerted over an ever-enlarging sphere, for every new acquisition of force and control increases the range of their operation, and affords fresh ideas and material for reflection. Activity entails a great variety of new experiences. Hence this may probably be regarded as the *troublesome* stage of infant life; for activity, without experience to guide it, is apt to become meddlesome and mischievous, even when quite innocent of intentional harm. Before baby could move about independently he was easily kept out of scrapes, but now he must be constantly watched, and his active tendencies must be either repressed or directed. Seeing that the use of the corporeal faculties is the appropriate way of strengthening them and of bringing them into effective use, activity should be rather encouraged and directed than repressed and thwarted. This is all the more necessary as the experience thereby acquired respecting the relations, qualities, and uses of things is most serviceable, and contributes a considerable addition to the mental stock.

Busy limbs are better engaged in some salutary employment than left to desultory and mischievous devices, especially as the thoughts will likewise be enlisted and kept from wandering and moodiness. A habit of attention may thus be initiated. To this end, attractive objects are wanted to captivate attention. The methods and appliances of the

Kindergarten system are very successful where a number of children of about the same age are brought together; but they do not answer well in the nursery, nor are they so applicable there. Of all the means ever devised to attract and maintain a continuous, untiring interest, a soft doll is, beyond all comparison, the best. It is not only the most effective means of captivating and keeping the attention, but it nourishes the best feelings and sympathies of the infant mind. It also affords the most useful and reliable indications of what is going on there. The Kindergarten "gifts" have not this tendency; they are too mechanical, and would have little enduring interest of themselves, apart from the suggestions of tutors. The old-fashioned, time-honoured toy of toys, the doll, is incomparably more attractive and useful, and inspires more permanent interest, than any other.

Imitation is a natural propensity, equally manifest in the infant as in the offspring of the lower animals. It is the natural means by which a large portion of the experience of infancy is acquired, especially before language is understood. It is generally most noticeable in peculiarities of gesture and manner and in mechanical habits; but it extends to conduct also, where the resemblance of the copy to the original is not so obvious. The lesson to be learnt by tutors from the close imitations frequently exhibited by children is to be on their guard, for it is not at all likely that pupils will always select for imitation that which tutors would most like to see reproduced.

Articulation is a special subject for imitative effort in infancy, and one which often taxes the powers strongly. A distinct utterance is an accomplishment seldom perfectly mastered before the very end of this period, and is generally far in arrear of the power of understanding spoken language.

The nervous system is not the least important part of the corporeal faculties at this period of life, when mental sensibilities are often so strongly excited, and when fear is so easily aroused. The utter inexperience of infants makes them easy victims of imposture, and their credulity is often practised upon to constrain obedience, to quell ill-humour, to divert attention, or to compass any trivial end. The most cruel and mischievous are those fictions which act on the nerves by creating fear. Ghosts, bogeys, and such-like nursery fictions are often fixed deeply into the infant mind. The terror thus wickedly aroused by ignorant nurses without the knowledge of parents, and even by some foolish parents themselves, is often productive of nervous disorders which last through life. Such are stammering, biting the nails, epilepsy, or even insanity. Long after all serious belief in such fictions has been discarded by reason, and when their origin has been forgotten, the terrors so inspired in infancy recur involuntarily in after-life in various forms of nervous affection.

Considerable difference of natural acuteness and power often exists between the eye and the ear in the same individual, and certain idiosyncrasies are due to that cause. Ideas formed more freely or more effectively through one or the other will differ accordingly. The habit originated by the preference of the more sensitive organ has an important bearing on the development of the mind, and should be studied, in order that treatment may be adapted accordingly. So little attention is generally given to this feature of infant sensibility that several years often elapse before defective or peculiar vision is detected. Defective hearing is more easily discovered, and therefore is not so often neglected.

OF THE WILL.—With the consciousness of power is ever associated the desire to exert it. Hence the period of infancy is a critical stage of mental history, when the faculties of body and mind begin to exercise a measure of independent action, and the Will becomes influential. Every increase of the faculties enlarges the means for the Will to assert itself, and adds at the same time both stimulus and opportunity for its assertion. Before language is understood, reason enthroned, or moral principle established, this powerful factor of the future character comes into force. And the manner of its exercise will naturally depend on the form it has received during the previous plastic stage of babyhood.

Power is a great demoralizer. This is not the place in which to discuss the passions its unrestrained sway may develop even in the noblest characters. These results are elsewhere mentioned, but the consequences of uncontrolled sway even in the periods of childhood suffice to show the paramount importance of giving a right bent to the action of the Will from the very earliest of its manifestations. It may then be moulded in such manner as to make it give strength and vigour to the higher nature on the one hand, or, on the other, to favour the predominance of the lower in the future character.

If the Will be neglected, and permitted to assert itself without due control, in the period of babyhood, the result will be shown in the too familiar spectacle of a spoiled child—one of the greatest of domestic nuisances. On the other hand, if the Will be suppressed by strong and decisive treatment—coerced, in fact—outward conformity and seeming obedience may be exacted. This, however, is not the exponent of the Will of the child, but of the force of the tutor. The child's own Will may be thus concealed for a time, but will probably betray its true character when restraint is relaxed or removed.

There is a time, a Divinely appointed time, when the Will is amenable to gentle suasion, when, in fact, it is virtually in the control of tutors; but if that opportunity be lost, it can never afterwards be fully retrieved. The work is then easy, and the response is natural and entirely free from the sting inseparable from sensible opposition.

OF THE MENTAL FACULTIES.—The dominant influence of activity applies equally to the mental as to the bodily powers. The perceptive faculty, sharpened by use, has now a readier power of apprehension than before, and the means of observation increase as the sphere in which it is exercised enlarges. External objects are more attractive now, and divide the attention which in the previous period was chiefly engrossed by internal sensations. The subject-matter of Thought embraces a larger proportion day by day of what is external to the infant himself. He is no longer a passive recipient of what presented itself unbidden to his senses. He now seeks for himself, and sometimes makes discoveries which, however useful as experiences, are not all agreeable either to himself or his tutors.

Personal sensibilities continue to operate powerfully as a stimulus of thought and action throughout this second stage, though they are less predominant than formerly. Neither the wants nor the physical sensibilities are so exigent, but mental susceptibilities become sensitive.

Children at this age are not treated with so much tenderness as in their babyhood; and they are conscious of the difference, especially when subjected to inconsiderate kinds of restraint and correction. Their feelings are not so much respected as before, and their activities involve them in frequent collisions with tutors. Thus their sensibilities are hurt in a new form; and a novel source of irritation is set up.

The treatment infants receive still supplies a considerable proportion of the subject-matter on which their mental faculties are exercised. The ideas so formulated are more definite than those of babyhood, though probably not so absorbing. Hence it behoves parents to know the kind of treatment to which their children are subjected in their absence. The scenes daily exhibited, in the streets and parks, of the proceedings of nurses, and of the conditions to which their charges are exposed, are very suggestive. Viewed by the light of natural law, which inevitably works out the natural result in every case, these scenes account for many strange tempers and other unaccountable developments in the infant mind.

The imitative propensity of infants may be seen in their conduct to dolly, which exhibits the workings of their minds and the ideas imbibed in their own experience. Dolly is a companion, and, above all, a pupil on whom are lavished all the attentions baby has himself received, and especially the persecutions and punishments he has suffered. Thus may tutors see how their own conduct has been observed, criticized, and understood (or mistaken) by their infant pupils.

Baby turns himself inside out in his confidences to dolly, and exhibits, by his imitation of the examples he has seen and suffered, the ideas he has formed. If the object of training were merely to impart knowledge of material Nature, the Kindergarten "gifts," coloured balls, bricks, &c., would be well adapted, with the mechanical ideas they suggest, to effect that object, but they do not encourage sympathetic emotions and moral principles. As we hold these to be of paramount importance, we regard the doll as a preferable plaything, because of the emotions it inspires and cherishes. The "gifts" may employ the hands and eyes, but dolly engages the whole nature, and especially the heart, the seat of love and sympathy.

Should a younger member have entered the nursery in the meantime, the comparison of present with past treatment is emphasized by the tenderness shown to the usurper. Happily for the one deposed, his activities are a constant resource, and if they get him into many scrapes, they also afford him effective diversion from many painful reflections.

External objects now occupy a good deal of attention. The organs of sense being by this time well practised and in full use, and the perceptive faculty quickened by exercise, the infant's powers of observation are much increased, and are actively employed in the enlarged sphere to which his newly acquired bodily powers introduce him.

This period is characterized by remarkable acuteness of observation, and affords opportunity for accumulating a stock of ideas for after-reflection. Nor is this latter mental process inactive, as will be seen hereafter; for, though subordinate, for the present, to external perceptions, it fulfils important functions in this period of general activity.

Through life, observation of the conduct and example of other persons is always more attractive and engaging than the aspects of inanimate things. Action and its results are more interesting than inert things and their qualities. And this rule obtains with special force in the particular period of life's history now under consideration. Infants, in common with the young of the lower animals, manifest remarkable acuteness of discernment. And this characteristic is most frequently and inconveniently displayed in the discoveries infants are continually making of things they were not intended to see or supposed to notice. These give rise to frequent collisions between them and their tutors. Whatever baby was not to see or to know, he seems to have a sort of intuition to find out and betray. His power of discernment is, in fact, far greater than that with which he is generally credited. The too common consequence is, that he is unjustly and unreasonably punished for the natural exercise of his natural powers. The imprudence and short-sightedness of tutors are visited upon him. The petty devices and frauds practised to screen things from him are transparent as glass to his keen perception; and he is accused for seeing through them. His sharp vision and acute hearing discern the intentions, ideas, and inner workings of the adult mind, as if by intuition, almost simultaneously with the ideas themselves. A sharp little fellow of our acquaintance, of four years old or less, used habitually and involuntarily to anticipate and tell what one was going to say before a word of it was uttered! And this habit, though sometimes very provoking, was so surprising that it was rather encouraged and shown off. We may here digress to remark that, in after-years, up to the age of thirty, when we lost sight of him, he showed no unusual power of perception or other mental attainment, but the extraordinary discernment he showed in childhood amounted to a sort of clairvoyance. The form of discernment in question is a characteristic of the period, and not of special individual power of mind. We lay some stress upon it because of its consequences in after-life. These arise from the immediate effects produced by this abnormal power of perception on the conduct of inconsiderate tutors who administer correction when it is undeserved; who visit the punishment due to their own fault, on the natural and blameless exercise of a natural faculty. And what makes such conduct more injurious in its effect is, that when this special acuteness happens to lead to amusing results, it is applauded, and perhaps rewarded. What can a child possibly think, feel, or infer from such unaccountable inconsistency in his exemplars? He *does* think of it, and is sensible of the *injustice* it involves.

The fact of this remarkable acuteness of perception as a characteristic of the period, though differing considerably in the degree in which it exists in different children, should suggest, as a rule of nursery conduct, that nothing should be done or said in baby's presence which he ought not to see or hear; nor should any of those petty deceits and frauds be permitted to be practised upon children which they would be punished for repeating by imitation.

It may be here incidentally remarked that children, even at this age, have generally very definite notions of injustice, and will not unfrequently remonstrate by gesture against unfair treatment, even before they can express it in words.

Memory.—Seeing that those ideas only which are remembered are reflected upon, and also that those will be most reflected upon which are uppermost in the mind, it follows that, for some time after the term of babyhood has expired, the ideas then formed will constitute the principal subject-matter for reflection whenever the mind is not engrossed by the activities of the period. The experiences of babyhood will therefore continue to be influential in this period, but attention will be gradually diverted from reflections on past experiences, and memory will be exercised in the great mental work of the period—the acquisition of language. The idea of significance which was already in babyhood beginning to be attached to certain sounds gains force continually, and soon becomes the chief occupation of the infant's mental activities.

Complex ideas involving reflective processes of *comparison, association,* and *inference* are now freely formulated, and show considerable power of discrimination. Reflective action, however, is chiefly such as is suggested by, and included in, observation, and is not in this period so much occupied with the past indeterminate experiences as with present mental activities. Things are now regarded with reference to their qualities and relations by comparison with former remembered presentations of the same. In this way knowledge of external objects is enlarged. Still, a considerable part of the subject-matter which the activities of the period suggest arises out of the consequences to which the corporeal activities necessarily lead. The independence of action yet permitted to the infant is very partial, and, unless his environment be carefully and judiciously adjusted, he will be continually checked and thwarted by tutors with scanty regard to the feelings of the little meddler. To be thwarted is not an agreeable feeling at any period of life, but is never so provoking as it is in infancy before the reasons which prompt it can be understood. Moreover, it contrasts strongly with the indulgence previously enjoyed, and excites strong feeling particularly trying to temper. Comparisons with the recent past are disturbing.

Happily, the infant's mind is diverted from the troubles into which his bodily activities betray him by the new impulse to understand the speech continually being addressed to him, and which involves the particular form of association which chiefly distinguishes and characterizes this period—that by which language is acquired; or, in other words, that by which ideas become associated with articulate sounds. This form of association, commencing in babyhood chiefly between gestures and their significance, is now exercised mainly between articulate sounds and their meanings. The impulse of imitation, one of the very earliest forms of activity, leads to early attempts to utter articulate words in the way of speech; but expression is necessarily acquired slowly, as it involves several factors which are not implicated in the mere understanding of the meanings and associations. The different processes are defined in a previous chapter, " On Language," whence it will be seen that the mental processes are by no means simple. Language is an achievement requiring a degree of effort not generally recognized. Children are supposed to acquire it naturally, or by a sort of intuition; but this mistake arises, like many others connected with the working of the infant mind, from want of observation of what is always going on there. The acquisition of language is effected by a constant habit of associating uttered words with connected actions, until the articulate words become representative of the inferred meanings. It is a laborious process, but is carried on under a natural impulse by continual unnoticed effort, and therefore seems to come without effort.

Inference is an essential factor in the acquiring of language. The names of things may be suggested by mere association, but the meanings of other words must be inferred from the manner and circumstances of their use. In such inferences the factors are few, and are presented to the mind simultaneously, wherefore they are of the simplest kind, and are not like judgments, the data or evidence for which are often numerous and wholly or in part supplied by the memory. Moreover, as the value of a judgment depends upon the sufficiency of the data or evidence on which it is formed, it is the process in which all the reflective faculties culminate. Hence it is the weak element of the juvenile mind, especially in its earliest stages. Even in the elementary form of inferences necessary for the acquisition of language, they are the weak factor of the process. Children rarely misapply names, but they often misuse other words.

Seeing that the infirmity of infant judgment is universal and its cause obvious, it should in every case receive careful cultivation by means of the method of correction explained in connection with Rule 6, chapter xvi. Such care is all the more necessary, seeing that children exert their feeble powers on whatsoever suggestion of personal feeling or external circumstance may happen to be at the time most strongly impressed or most obtrusively present in their minds. Of whatever nature may be the subject-matter uppermost, whether it be suggested by the conduct of tutors or by any other kind of circumstance, whether of the lower or the higher nature, sensuous, mental, or moral, such also will be the character of the reflections and the ideas into which they enter. This is at once indisputable and obvious, yet the fact is practically disregarded in reference to the subject-matter of infant Thought as provided for them or subjected to their contemplation. Seeing that infants have little or no choice in the matter, and *must* think of what is actually presented to their minds by the environment, it follows that, until they acquire independence, their minds must be nourished by whatsoever aliment is so supplied to them. No wonder that they sometimes develop unexpected tendencies, when the first processes of formation of character were determined by chance, and whilst they were supposed to be impervious to Thought and innocent of any idea of inference!

Locke is hard upon infants. He devotes but a few lines to infant Thought in his great work on the Understanding; but he opines that children are generally spoiled by overmuch "cockering," and should be taught self-denial. We freely grant that fondling should be *consistent*, not fitful; and that it should be given for encouragement and for inspiring love, not to smother ill-humours or to pamper a crooked Will. We would have none of such cockering. Respect for their personal sensibilities, however, we hold to be in no proper sense deserving that name, and, on the other hand, we plead Locke himself in favour of such tender regard, for he maintains that *judgment* necessarily proceeds upon the data present to the mind ; and he alleges that, if all men saw the same things in the same light, all would arrive at the same judgment respecting them. In other words, every mind, whether young or old, educated or uneducated, human or brute, will judge according to the ideas present to his understanding. Certain it is, that the subject-matter supplied to the infant by his tutors and surroundings will naturally determine what he thinks about and how he judges.

In this period the acute discernment of infants is often employed in very smart criticisms and judgments on the conduct of tutors, who need to remember this natural tendency, and will not do well to despise it.

TEMPER, originated during the helplessness of the previous period, becomes more demonstrative and decided as the powers of exhibiting it increase and opportunities multiply. Treatment at this stage is better understood than it was before, and is therefore more influential, for it suggests other ideas beside mere feelings : such as reason and justice. If the bent given previously be good, the trials of this period may be withstood, and good temper confirmed ; but if it be otherwise, the crosses and thwartings of infancy will almost certainly make it worse. In the natural indulgence of his natural activities, the infant is sure to get into mischief; which means that he will find, and perhaps destroy, things which ought not to have been in his reach. Then he will probably be punished unjustly and irritated sorely.

Sympathy, aid, and interest in his proceedings, on the part of tutors, will be well appreciated, and produce their proper fruits—love and trust. The want of such consideration will as naturally inspire repugnance or indifference, as the case may be.

FAITH.—Faith is, in the first period, almost implicit, and so continues during the time of entire dependence. Afterwards, it can be an actuating principle only in so far as it is inspired or attracted. Hence, in this second period it may either be established or dissipated. In regard to this vital element of character, this is a critical stage of the personal history. The conduct of tutors should therefore be specially directed to this point. *Consistency* is essential to success. Considerate regard and sympathy, a manifest interest in, and entering into, their proceedings, will attract the confidence of infants if the trustfulness natural to babes have not been already shaken. The absence of these qualities will as certainly produce an opposite effect.

The natural effect of confidence is a desire to please—a voluntary yielding to those in whom it is reposed. St. Paul applied a natural law to a spiritual doctrine when he declared that it was impossible, without faith, to please God. Faith is not a religious dogma; it is a universal law in mental as well as in moral teaching.

As the actuating principle of true obedience, it is of paramount importance to the future success of tutors and to the progress of pupils. Without it, the relations of tutor and pupil are out of joint; the tutor must then rule by force, and the pupil be in subjection. The one must drive, the other be driven. Their *natural* relation is quite the reverse of this:

the one then leads, and the other follows by his own act of choice, attracted by the natural impulse of faith.

Faith opens the door of the heart, and lays bare its inner workings; but distrust closes up all the avenues to the heart, and begets concealment, which is equally inimical to personal peace and to friendly relations with neighbours. Moreover, distrust reacts, for it excites distrust in others, and the secretive habit it engenders dissipates their confidence.

OBEDIENCE.—Unless established on its own natural basis during this period, there is much doubt whether it will afterwards be acquired. It is generally so much easier to gain obedience by means of force and fear, that tutors too often have recourse thereto, rather than pursue the wiser but more difficult course of winning the confidence of their pupils.

Parental authority is the basis of faith and filial obedience in the natural family, as that of the Almighty Father is in the spiritual world. Hence, duty in either case is a joy and satisfaction on the one hand, or a task and a trouble on the other, according as obedience is the spontaneous offspring of faith or the enforced compliance of superior power.

The prominence given to the commandment to children to honour their parents proves how important a principle is involved in the recognition of authority through the medium and under the influence of natural affection, as opposed to its enforcement by constraint.

Obedience, if duly observed, reduces to a minimum most of the conflicts with authority which the activities and inexperience of infancy entail. It should therefore be with the infant, as it was with Adam when he became invested with a moral nature, the first lesson. As faith in God is the basis of true virtue, and as obedience is its first-fruits and exponent, faith in the authority of tutors is the parent of obedience in children, and is the first principle to be established thereby. When Adam had to learn obedience, the Almighty Father took His own wise way of teaching it. Had the consequence of disobedience been at once realized, there would have been an end of the lesson and its purpose. The forbidden fruit was good for food, and to make one wise. The impulse to eat it was natural, and not in itself wrong. The fault was in the want of faith, and of the consequent want of obedience to the commandment. These were the lessons Adam had to learn, and which all his posterity have also to be taught. In Eden, Adam needed no other kind of knowledge than what the lower animals possess. He was thence driven out to a sphere where good and evil were manifested, and where obedience, founded on faith, might be acquired. When man has fully learned that lesson, he is fitted for a far higher destiny than Eden.* Life is the time and opportunity for learning these lessons of truest wisdom; and infancy is the time when it is first brought into requisition. Obedience is a necessary qualification for the conduct of life; and is most easily made intelligible to infants, and reduced to practice in all periods of life, by the influence of faith in authority.

MORAL PRINCIPLES.—These will still be inferred from, and suggested by, the example of tutors rather than by direct tuition. The principles involved are

---

* See " Good and Evil," by the same author.

practical, and cannot, in fact, be so well expressed in language, even where a full knowledge of language exists, as they may be exemplified in practice. Dolly is a most influential medium for moral principles at this period, and here attains its highest value as a means of instruction to both tutor and pupil. The latter learns the nature of moral obligations, and the former learns the state of his pupil's mind in reference thereto.

The ideas of moral obligation imbibed in infancy are amongst the most effective and enduring of all the acquirements of the period. Consistency on the part of tutors is, however, of paramount importance, for the acute discernment of infants, acquired by observation before the powers of perception are diverted to speech and other mental effort, will surely be exerted in criticizing their conduct, and will not fail to detect inconsistency. The common habit in the nursery of having one rule for the children and another for tutors is destructive of true moral principle. It reduces moral obligation to a very low order of expediency, to be assumed or doffed as convenience may dictate.

Of all moral principles, *truth* is that which is most important to be observed, and, unhappily, the one most frequently infringed by tutors in their conduct towards infants. The little duplicities and deceits practised upon them, the persuasive glosses by which they are beguiled, and the cunning artifices by which truth is so often veiled or distorted—all the multiform compromises by which the sterling principle is evaded—go to form the ideas which are always in process of formation in the infant mind. Artifice, however skilfully devised and ingeniously executed, is almost always discovered sooner or later by the acute discernment of young children, and is a fatally effective means of instilling guile. If those susceptible tablets could be exposed to view, and exhibit the deeply graven lessons in guile that were inscribed there even before the child had yet acquired the power of expressing his ideas, how few tutors would read those characters without surprise and sorrow! Thus it too often happens that, by the time he has learnt to speak, the infant is already an adept in the art of lying, especially in that most deceitful form of falsehood, white lies—lies that simulate the truth.

In common nursery practice there is probably no other influence so injurious, nor any so general, as untruthfulness in all its appalling varieties. No idea is more readily apprehended by the infant mind than deceit. We do not allude to downright falsehood, which is abhorrent to all but the most depraved minds, but to prevarication, excuses, misleading remarks, and such-like infractions of the true principle. It is unfortunately no uncommon thing, therefore, to see an infant of three or four years old put into requisition all the arts of evasion and duplicity to conceal a fault. Such depravity has not so remote an origin as that to which it is commonly attributed.

During this period children are very commonly taught hymns and pieces of poetry, which they learn to recite with such accurately imitated intonation and accent as to inspire the belief that they understand the sentiments expressed. The cases in which some general idea of the meaning of these recitations is comprehended are, however, exceptional. As a rule, they are mere imitative efforts, like the utterances of parrots. They are certainly not to be relied upon as evidence of knowledge. The religious doctrines intended to be thereby taught are quite beyond the capacities of infants, both in regard to the ideas themselves and the terms in which they are embodied. As exercises of memory they are of doubtful value; and as a means of instruction useless until language is fully understood. Considering the pains required both in teaching and learning them, we think such exercises were better deferred to a later period, when language is better understood and the understanding is more mature.

HABIT OF THOUGHT.—The habitual state of mind towards persons generally is more or less definitely formed during babyhood, and is one of the most important fruits of that period. Other habits thereafter acquired will be influenced thereby. Habits of Thought may be induced unobserved, by relative qualities of the eye and ear—by physical circumstances, and by conduct of tutors; but, howsoever induced, they will be determined mainly by the subject-matter on which the thinking faculty is exercised at the time the bias is given and the particular habit is initiated.

During this period of activity, certain gestures, uses of the limbs and faculties, will naturally become habitual. All indications should be watched. Good tendencies may thus be encouraged and confirmed, and others diverted or corrected, in order that this powerful factor of future mental progress may be enlisted in right directions before it becomes too strongly implicated in wrong ones.

Attention (the most useful form of control) and obedience should be cultivated carefully at this time;

for if they be not established during infancy, it is probable that their opposites will have become so in greater or less degree. The balance will not stand at zero; it will turn one way or the other. The Kindergarten exercises are eminently useful in attracting attention, and in initiating a habit of directing Thought to a given subject. To us this appears to be their most useful and practical result. Interest is the only *effective* means of attracting and sustaining attention. It follows that attempts to enforce attention to matters possessing no interest for the child will inspire dislike of the lesson so imposed, and thus defeat its purpose. Without interest there can be no effective attention.

SUMMARY.—The progress made during infancy, the evidence of the work done by means of the first individual activities of body and mind, is represented as follows:—

1. By certain physical development, including effective use and control of the corporeal faculties, some dexterity of limb, and fair power of articulation, and probably some influence, for good or evil, on the nervous system.

2. Increased power and facility of perception, which has been exerted over a larger environment, and has resulted in a stock of definite ideas of external things, their relations, qualities, and uses.

3. Decreased sensitiveness of physical feelings, and a corresponding increase of mental sensibility.

4. Both the power of understanding speech and of expression in words have been acquired, and also a certain knowledge of the construction of sentences. These acquisitions imply a continuous mental effort. They mean that a great number and variety of com-

mon objects, their qualities and relations, and also a great variety of actions, have been definitely associated with representative articulate sounds, in such manner that they are appropriately and freely employed in speech. This accomplishment involves strong effort of memory, association, and other reflective faculties.

5. The formulative faculty has been diverted in part from personal feelings by the superior attraction of external objects, and has found ample employment in the acquisition of language, and in connection with the novelties of independent activity. The elements of graphic language may also have employed the mind. The subject-matter of Thought suggested by the new activities has been divided between ideas of personal and external interest, with mental results corresponding thereto.

6. Consciousness of powers, bodily and mental, which gave the impulse to their active use, has strengthened the Will in whatsoever direction the first bent and subsequent management may have imparted to it. The child is more or less either spoiled or tractable, headstrong or amenable, stubborn or docile, at the end of this term of his natural history.

7. The attitude or relation of the child to other persons generally, and to tutors in particular, will accord with the feelings and experiences in which that relation originates. For instance, continual thwarting and irritating correction and interference could not make its subject amiable. Attraction and repulsion follow the same natural law in intelligent beings of all ages and of whatever type. Children so treated may sometimes be *made* to *seem* otherwise

than as they feel, but such semblance is far different from the true principle.

Love, confidence, and obedience, or their opposites, will have been established inside, whatever the outside covering may be, before the end of this eventful period of the mental history.

Moral principles, too, will have taken root according to the sowing. The harvest will not be referable to Adam, but to some of his very latest posterity, especially in relation to that most fundamental principle, *truth*.

The rudiments of the future character are definitely formed, not of course unalterably, but firmly, in infancy.

CHAPTER VIII.

ON THE NATURAL HISTORY OF THOUGHT.

THIRD PERIOD: CHILDHOOD.

THIS period commences when the free use of speech has been acquired, and it terminates with the acquisition of written language. It is the preliminary to school-work, and supplies the necessary equipment for education in the technical acceptation of that most important work.

The first of our periods commenced with nothing but sensibilities to direct the incipient faculties, which therefore dominated till active powers were attained. These were the ruling influence of the second of our periods, in the course of which speech was acquired. This, therefore, is the new element of the equipment to direct the proceedings of the third of our periods. On the natural principle that conscious possession of a power prompts its use, speech is the dominating influence of childhood, and *inquisitiveness* is its natural characteristic. If the child be not a chatterbox when he is in full possession of speech, he is not quite in a natural condition.

Speech gives a new direction to the mental impulses. Up to this time the infant has been industriously *observant*. He has thus acquired a very extensive knowledge of the *aspects* of things.

By reflection and active experience he has also obtained some knowledge of their qualities, relations, and uses. Now, however, he is not satisfied with this superficial view, and he is inquisitive as to the reasons why and how all these things are so. Speech has imparted a new interest to all his activities and observations, which begin now to be subordinated to the higher purpose of *reason*.

If tutors recognize this natural turn of affairs in its true significance, they will see how well it befits the situation, how wisely it has been ordained. It will be apparent that, in the existing condition of the child's mental furniture, aspects of things are no longer sufficient for his future purposes in life. *Reasons* must be added, in order that reason may assume its proper place and functions in the future character and conduct. If tutors fail to see in the child's inquisitiveness a natural and appropriate condition of the period, they will lose their opportunity and the child his advantage. If they snub and repress inquiries prompted by appetite for knowledge, and try to give *other* knowledge for which there is no appetite, they will utterly fail of their object and defeat his, thus creating mutual irritation and evil. If they commit the common mistake of wanting him to put away childish questionings whilst he is still a child, they will but contravene Nature and kick against the pricks. If they refuse to satisfy his natural appetite, and try to create instead some other more to *their* mind, they will encounter the greatest difficulty teachers have to meet, and they will fail. They will need the child's *attention*, and that cannot be forced. Instead of availing themselves of the attention he volunteers, and using the opportunity thus afforded, they generally try to divert it, in doing which it is dissipated. Instead of giving the food he craves and could digest, they endeavour to substitute something else, which he does not want and will not have. Teaching is thus made repulsive, and lessons are converted into tasks before the proper time for either, and when both might be made a pleasure.

OF THE CORPOREAL FACULTIES.—Increased knowledge of the outward aspects of things leads to inquiry as to their qualities, uses, and particular relations. The corporeal faculties are therefore employed in investigation of things. The environment is subjected to prying research. Limbs lend ready obedience in the work of the period, and possess aptitude and power by virtue of previous practice.

Senses are acute, practised, and discriminating. The activity of the former period is turned to the scrutinizing purposes of the present, and large accessions of knowledge of persons, character, and things result.

Occupation for both body and mind, primarily for the bodily powers, has a salutary effect on both, and should be devised to engage the interest and suit the taste of children. It will then be equally advantageous for the relief of tutors and for the welfare of their pupils. It should be useful, in order that it may not teach the waste of time—the stuff life is made of. On the other hand, it must by all means be made agreeable. Whatever tends to make work a *task* is most injurious both in present influence and future tendency. Seeing that this is the term for the preparation for school-life and for the work of regular education, it would be a fatal error to inspire a distaste for work. This is a question of care and tact in the management of children during this period of their natural history, and is worthy of the best attention of tutors. Their choice is between the two natural forces—attraction and repulsion. There is no other alternative.

The exercise necessary to health and to the development of the bodily powers will lead to an enlarged environment, companionships, and influences which will materially affect the present condition and future character. These, if not the objects of tutors' solicitude, may entail serious results on the future mental condition.

A useful expedient, commonly employed in the lower grades of society, is to invest elder children with a sort of protective charge of the younger. Where this can be done, it is in many ways useful. It affords much the same indications of character and opportunities for training the elder that his treatment of his doll afforded in an earlier stage of his infancy. A further advantage is, that it suggests a sense of responsibility, inspires an idea of duty, and prepares the way for higher charges in after-life. It brings out some at least of the effects of previous training; and, as a rule, tends to develop the higher rather than the lower nature.

OF THE WILL.—The management or neglect of the Will in the previous periods will have determined its force and character in this. This stage of the history will have commenced with a Will already formed. Its present bias may be controlled by the restraints of domestic discipline, but its further modification as a principle will depend upon example, affectionate remonstrance, appeals to reason, and such-like means. Still, the force of habit acquired before these considerations could be urged with effect, or be even understood, will be most influential.

The condition of the Will at the beginning of childhood will have a marked influence on the nature and extent of the mental progress of the period, for work done willingly is more effective than that which is done reluctantly or under pressure. The tendency of force is to excite resistance; and hence work done under compulsion is done under a tacit sense of resistance, and is correspondingly ineffective. Like food taken without appetite forced lessons lack the condition necessary to effective digestion.

OF THE MENTAL FACULTIES.—Perception is now almost mechanical, except in regard to things new or strange. Effort, therefore, is chiefly devoted to formulative and reflective work. Corporeal sensibilities decrease, but mental sensibilities occupy a large share of attention and form the subject-matter of much reflection. The conduct and example of tutors and others are critically considered during this period, and exercise much of the inquisitive power. Judgments thereupon will naturally be according to such standards as may have been established by the relations previously formed.

The mental work proper to this period is the natural result of the dominating influence, inquiry, and will consist of *learning* as distinguished from *observation*—of knowledge received from others, rather than of what is derived from his own perceptions. It is the special function of language thus to extend knowledge, and concentrate general experience to the use of the individual. In former stages the child had to get what he could for himself. Now he receives the fruits of others' labours and experiences, condensed into verbal formulæ. By means of language he sees not only with his own eyes, but with the eyes of many generations.

Reading is at first practised without the sense being understood. The process of spelling the words, letter by letter, is early dispensed with, but, long after that has become unnecessary, the effort of grasping the words individually continues for some time to absorb so much attention that even their relative places in the sentences are not apprehended. And when a facility of grasp sufficient for the mastery of these two elements has been acquired, the sense or meaning of the whole in its connection has still to be perceived. Hence the early efforts of reading involve a mastery of the literal and verbal combinations rather than a gathering of knowledge therefrom. Attention is necessary both for deciphering and comprehending written language. Hence, until great facility has been acquired by practice for the former, there is none to spare for the latter. Tutors have need to make allowance for this. A child of five, who was fond of reading, always chose from his father's library "Satan's Devices," which, of course, he could not understand. Other books more suitable were often substituted by his tutors, but his preference remained unchanged. The truth was that all were equally unintelligible, and his choice was probably due to the clear type and handy size of his favourite. In this connection, and as elucidating the double action involved in reading, it may be here remarked that print is much more readily understood than manuscript, even by professed readers, because it is more easily read.

Personal feelings, though different in kind from what they were in babyhood, are not less influential. Corporeal sensibilities decrease year by year as physical nature becomes stronger and more developed, and also because superior forces attract the thoughts to other channels. But whilst the physical susceptibilities decrease, mental sensitiveness still more effectively increases. Hence the storing of ideas proceeds continuously from first to last; but as the nature of the subject-matter on which the faculties are exercised changes from time to time, the nature of the thoughts produced differs accordingly.

It would be quite impossible to understand child-nature without a thorough appreciation of the personal feelings proper to that period. As a rule, they are but little regarded, and are therefore often deeply wounded for want of a due recognition of their sensitiveness and a proper respect for their claims to consideration. We allude to the unceremonious way in which children are put aside, repressed, thwarted, and in innumerable ways habitually humiliated. The ideas naturally produced by this sort of treatment are of a very definite nature, and form a very considerable portion of the currency of mental reflection during childhood. Their influence is necessarily very potent, being proportionate to the sensibilities so excited; and it is also, by natural law, of the same kind as its parent cause. As is the sowing so also will be the reaping.

The idea of showing politeness and respect to a child seems to most people as too ridiculous. A strong-handed disciplinarian would protest against having children "stuck up" and made to think too much of themselves; but, though assuming, without evidence, that such effect would be produced, the natural effect of the other course is not duly considered. The natural and inevitable consequence of an example of conduct the very opposite of what is required of the child himself sets up one rule for the tutor and another, its opposite, for the child, and makes *precept* an empty name. The precept enforced on the child is in such case continually violated by the preceptor's own example. We deny the assumed consequence of considerate treatment, and confidently affirm, on the highest authority, that the golden rule is of universal application. The result of discipline will accord, not with theoretic intentions, but with the means used to enforce it.

At this period of life, when respect for authority in general, and faith in tutors in particular, is so important to the mental comfort and harmony of tutor and pupil, consistency between conduct and precept is a primary consideration. Respect for personal feelings contributes through life to the maintenance of mutual good relations, and never more so than between tutors and children at the age now in question.

Seeing that the prevailing force of moral principle as the rule of conduct is first inspired by, and is dependent upon its exemplification by, tutors, the consistency of their example with the precepts they inculcate will go far to determine the ideas of moral principle which children imbibe. Personal sensibilities are with them a practical test.

External objects are regarded with ever-increasing attention and effect as the mental powers acquire vigour by exercise. Niceties of distinction and minute particulars, either not perceived before or not appreciated, now engage attention and invest external objects with new interest. The uses, properties, and structure of things occupy much Thought, and supply material for reflection, calculation, and inference, at this inquisitive period.

The acutenesss of discernment by mere observation which characterizes the earlier period gives place slowly to the more regular exercise of reason, and almost disappears when the reasoning powers acquire vigour. The rationale of this appears to be that the higher supersede the lower forms of perception.

The character, motives, and conduct of others are especially subjected to scrutiny, and are discussed amongst children with much freedom and some acumen. Their criticisms prove that they are very observant, and that the treatment they receive is very influential in giving tone and colouring to the current of their thoughts.

The ideas now formulated are more definite and more deeply fixed in mind than those of former periods. Events especially of the latter part of childhood are therefore remembered through life; not as mere impressions, indeterminately, but circumstantially, with the particulars of places, persons, and other associations, thus proving strong original formulation.

Great progress is made in the use of language during childhood. Meanings are more definite, and words are daily invested with more distinctness of signification. The various processes involved in the analysis of articulate sounds and their representative graphic symbols are pursued with more or less interest and success according as they are understood or merely learned by rote. Reading and writing are sometimes made interesting pursuits, but are more generally converted into wearisome tasks by the manner of teaching. When this occurs, it is usually because the first lessons were given by unqualified teachers. Any one is supposed to be able to teach a child, but this is a great mistake. First lessons should be made as plain and as agreeable as possible, which requires tact and skill. There is in child-life a very great deal of learning quite independent of any teaching, and, on the other hand, there is a vast deal of futile teaching with no real learning. This subject is treated more fully in the sequel, and is illustrated in the case cited in Note 4 of the Appendix.

The relations of number and magnitude form the subject of attention, which results in some elementary ideas of arithmetic. These ideas are often confused and imperfect for want of a thorough grounding in the principle of notation. The relation of the values of figures with the places they occupy is seldom made clear to beginners, who therefore follow rules, or misapply them, without proper ideas of their significance.

The characteristic inquisitiveness of children affords tutors ample opportunity for teaching them a great deal, on the natural principle of letting them learn what they want to know, and in their own way. Their knowledge of qualities, relations, and principles not in themselves apparent may thus be increased indefinitely, and to their lasting advantage.

Memory.—Seeing that every idea which requires any reflective action, such as comparison, association, or inference, in its formulation depends upon the memory for the recall of the ideas involved, the stock of ideas necessary to meet this demand must be very considerable even to serve for the reflections common to all intelligent children of this period. The accumulation of such a stock must also have necessitated considerable mental work. It would be hard to estimate the stock, with any pretence to accuracy, at any given time, but some approximate idea may be formed from the number of ideas and operations involved in language alone.

As the meaning of every single word that is used must have been derived by inference from the connection in which it was heard, and that several, and in some cases many, repetitions of the word and the act of inference must have been required to fix it correctly, the number of recollections, associations, and inferences required for each word must have been considerable. Now, children at the age of five years ordinarily use about 540 out of about 1080 words which they understand (see Appendix), but we can only take account of those they use, the others being uncertain. It follows that 540 words and an equal number of meanings—in all, 1080 ideas—have been definitely formulated and firmly, almost inseparably associated together. The number of separate inferences required for each, corrections, questionings, and doubts included, would swell the number of mental acts to a very formidable amount. But large as the number is, it understates the actual mental performance, for it refers to language only, and to the words expressed, without accounting for those which are understood, but not as yet pressed into active service. Moreover, many of the meanings are themselves complex, such, for instance, as the common verbs *to think*, *to move*, *to do*, each of which comprises a considerable area of meaning. Yet, as these words are used appropriately, the meaning of each in its breadth must have been correctly formulated.

Some idea may thus be formed of the total number of mental acts involved in the mastery actually acquired by average children at the age of five years in this one department of their mental work. The proof that it has actually been done is attested by the fact that children use these words freely, apply them properly, and therefore understand them thoroughly. In the interest of children, whose cause we plead, we submit that this is a great work; and it proves more than is at first apparent. For be it observed, they do this entirely by themselves without being taught.

Now, considering the magnitude of the work, and the effective manner in which it has been accomplished, it is well worth considering *how* it has been done. That is a secret which the child cannot tell, and one for which our own personal experience does not fully avail us. Still, a good deal may be safely inferred. For instance, the work must always have been voluntary. The child drank in the ideas of his own accord. He was not crammed with them. The work was done quietly and unobserved. While the ideas were gently flowing and filtering in his mind, there was not a pair of expectant, impatient eyes fixed upon him. He was not disturbed by a running commentary on his incapacity, inattention, and other such scatterbrain influences. The full measure of his limited powers was employed without distraction. When he missed his mark, he was not accused, and had no fear of consequences. The mental operation was, like growth, slow, gentle, silent, continuous, and the effect sure, cumulative, and permanent. He learnt, as the lower animals learn, by example, imitation, and inference, without fear of punishment or reproach. Both the process and the manner of its performance were natural, not artificial. It was not done against time. It was not task-work. Furthermore, the child will go on, so far as his own proper language is concerned, adding to his stock of words in the same quiet way, and will perfect the work so pleasantly and effectively begun. He will gather the inferred meanings of new words with greater ease and with fewer repetitions.

Nor is the facility with which children acquire language peculiar to their *own* language. They will learn a foreign language and many other things as easily, provided it be done in the same manner. We have known a child of five years old speak in his imperfect way three languages. His father was a French-speaking Swiss. His mother was English. He had been brought up in Switzerland the first three or four years of his life, and was afterwards in Italy, under an Italian nurse. If addressed in English, he would reply, and proceed to talk in that language. In the midst of a game of play, we have adopted another language without warning, and he seemed to be unconscious of the change, following with the new tongue, and continuing it so long as it continued to be spoken to him.

The lesson to be learned from the foregoing remarks is that the conditions favourable for mental work are: a desire to know, freedom from pressure of every kind, and from care, pre-occupation, and distraction. The most unfavourable conditions on the other hand are those which are generally adopted by inexpert teachers who are not qualified for the delicate task of giving the first lessons. The common practice of delegating early tuition to unsuitable persons is responsible for much of the distaste for learning which is so often acquired by children, and which is so formidable an impediment to school-work.

Comparison and association, and other reflective processes, are effected, even at this early age, with so much facility and quickness as to appear to be simultaneous with observation, and not to involve consciously separate acts. The several simple relations are apprehended at once, in the same way that words are grasped in their entirety without the need of spelling, or separate consideration of the component elements. As an expert sees at once how all the pieces on a chessboard are related to each other, without requiring to note the position of each separately, so, in like manner, the mind becomes so practised in observation that the reflective acts by which the relations and qualities of things are discerned become mechanical, and are performed without conscious effort. Ideas are no longer apprehended singly, but in groups. Things are perceived along with their qualities in their own complexity, and are so formulated by a single act. Hence, ideas now formulated embody more complete and accurate knowledge than those of a former period, and are therefore more definitely stored and remembered.

The degree of facility with which association of ideas is mentally effected is a principal factor in the acquisition of written language, which is the chief and most important mental work of this period. In the act of reading, the ideas familiar to the ear have to be apprehended by the eye. At first, and for a considerable time, these operations are not simultaneous. A distinct act of translation has to be performed mentally. In speech, the meanings of the words are associated with the articulate sounds, but in reading these meanings must be associated with the *written* words. By speech, an effective partnership was established between the sound and the meaning; but now another partner has to be admitted, and he is a stranger belonging to another family. Still, until the new partner is fully indoctrinated into the whole business of the firm, each has to act separately. Time is required, as well as practice, for the new work of translating and associating ideas, and this is not abridged by the impatience generally manifested by teachers when they do not consider what the beginner has to do. Charges of inattention when the effort is going on do not assist him, but, in fact, interpose the greatest obstacle. Embarrassment is the most adverse condition for mental work.

Notwithstanding the proverbial quickness of practised Thought, time is always an important factor in mental processes. Thought is much more quickly effected by some persons than by others. Some are slow in perception, reflection, and communication; others are so in one or another of these factors. This fact is regularly recognized in astronomical observations, and is subject to special estimation. Every observer has his own personal equation, which enters into all calculations in which accuracy is required in connecting a given phenomenon with the time of its occurrence. This being so in adults, so thoroughly practised, how much more must it affect our little clients!

Language, when fully acquired, is a great help in mental processes, as many complex ideas are condensed into a single word, or into a brief formula of words. Still, language must have become very familiar before it is available as a vehicle of Thought. As power of expression increases with the means language affords, the knowledge of language itself becomes extended, and the connection between ideas and language expressive of them is so intimate that the two become inseparably associated. Then it is that language fulfils its chief use in mental processes, as well as its other purposes of oral intercourse and written records.

Inference and judgment are now in constant and practical exercise, and are influential in their effect on conduct, as well as in mental work. In proportion as conduct is independent, it indicates the nature of judgments formed, and, as these are determined by the prevailing currents of Thought and power of the faculties, conduct is a criterion of the state of mind and effect of discipline. In so far as conduct is the natural and spontaneous result of training, and is not dictated by artificial restraint, it shows the practical value of the knowledge acquired and the force of the principles imbibed. As each successive stage makes the child more independent, each in turn prepares the way for that entire self-dependence he will have to exercise when he embarks in life. Hence the curb should be used as sparingly as possible, lest it lead to concealment of the real tendencies, which, in such case, will assert themselves in after-periods, when correction may be difficult, or too late, to be applied.

Judgment, being the highest function, is that to

which all training of the intellect should be ultimately directed. If tutors have used wisely the opportunities afforded during the earlier periods, the judgments of the later will be less superficial and more reasonable. But if those opportunities have been neglected, judgments will afterwards be random, immature, and erroneous. It is for the purpose of forming correct conclusions and opinions that the other mental faculties are most needed. Accurate and complete ideas are the fundamental factors or evidence on which judgments and opinions are based. Memory serves to recall them as required, and concentration fixes the mental powers on the work of estimating the data. All combine in the resultant act of judgment.

The inquisitive tendencies which characterize this period of child-life, prompted at first by the impulse to employ the newly acquired power of speech, are well adapted to promote mental progress by directing the attention to ascertaining those qualities and relations of things which are not apparent. It is the appropriate supplement to observation. It is the best use to which language could first be applied, for the knowledge so acquired is precisely that which is required for sound judgments. Without such knowledge, these would be superficial, based on the outward aspects of things, which are so often deceptive. The subject-matter of thought, of whatever kind it may be, and whether relating to persons or things, is thus scrutinized and fitted for the purposes of just inference and sound judgment.

TEMPER, as a habit of mind, has been decided in previous periods; not that it is incapable of modification, but that it is definitely formed. It is always more or less amenable to the same influences in which it originated. It may be improved or otherwise modified, like other habits; but it is now become habitual and decided. Most other habits have a restricted operation, but temper either sweetens or embitters the whole nature.

At this period, example, being better understood, and being judged by its actuating principles, is more influential than before. Reason, too, is available as a resource and argument, and may be used with good effect. Still, as the ideas in which temper originated were indeterminate and general, they cannot be met by definite means. A definite argument may be met by a definite answer; but an indeterminate one can only be counteracted by experience.

The common practice of teazing children, especially those of an excitable temperament, is highly injurious and reprehensible. The natural effects of such treatment are illustrated in Note 1 of the Appendix. Such extreme results may not often occur, but they prove the nature of the effect produced, whatever the degree may be.

FAITH also is, in its inception, based on indeterminate ideas, but, being a particular relation and not general, towards certain persons and not to the world at large, it may be shaken, or even destroyed, by an act of injustice, treachery, or falsehood on the part of those in whom it was reposed. If firmly established in former periods, it will be a powerful resource to tutors in the present. Not only is it much easier and pleasanter to lead than to drive, but it is far more effectual. Driving is practicable only within narrow limits, but leading may be carried on to the full extent of the mental powers.

Hence, to maintain the confidence of pupils is always important, but is especially so when they begin to exercise their powers independently.

Besides, faith, as a general principle, as a recognition of authority generally, and not in its particular and personal aspect, is so dependent on the cultivation it receives in the lower domain that, if this be uprooted, the other is likely to be seriously shaken. If, therefore, faith in tutors be impaired and shaken, faith in God is thereby endangered.

During the trials incident to this inquisitive stage of child-life in the maintenance of discipline, the relation between the pupil and his tutors is necessarily strained, even under the most judicious management. Thus the faith of the former is severely tried. In this trial, consistency of example and conduct is far more influential than a high model. Nothing is more destructive of the confidence of children than that elasticity of principle by which tutors change it to suit particular persons and occasions. The relation between authority and its subject at this age affords good opportunity for the exemplification of moral principle in the one and its appreciation by the other. The result in most cases will be either a firm establishment or a decided loosening of confidence in tutors, and, through them, in authority generally.

OBEDIENCE.—The practical observance of obedience and submission to authority by children at this age is necessary in the family and elsewhere. It is therefore generally enforced where it is not voluntarily yielded. The moral influence of this period is largely dependent on the kind of submission which prevails. If it be the fruit of trust and confidence, inspired by a sense of duty, well; but if the submission be rendered under fear of consequences and restraint of force, when in fact it is only an outward appearance concealing a spirit of rebellion, friction is inevitable, and the moral consequences are adverse. This state of affairs may always be referred, and may sometimes be distinctly traced, to treatment in previous periods, such as unjust punishments, unreasonable conduct, provocations, and petty oppressions. These may be, and often are, unknown to parents, who then refer the obliquity of their children's tempers to hereditary or constitutional defect.

Lessons in some form are an almost necessary part of the discipline of this period. The conditions favourable to successful learning have already been specified, as well as those which are adverse and tend to inspire a distaste for tuition. The difficulty of inspiring interest in lessons and in attracting attention to them is often very considerable, and seems to imply disobedience; still, these conditions being indispensable, any attempt to teach in their absence will be even more troublesome, and at best but very partially effectual. In our experience, actual dislike has never been encountered except where it had been created by injudicious attempts to enforce lessons or where the mental powers were conspicuously deficient. Stupids in general are not so from natural incapacity, but from mismanagement in some form. For instance, we were assured, with regard to a child of twelve, that it was quite impossible to make him comprehend the simplest rule of arithmetic. He was alleged to be incorrigibly stupid at figures. At first he seemed to verify that character, but after some patience we found that he had no proper idea of the principles of notation. His familiarity with the set formula of tens, hundreds, and thousands gave the idea that he knew their significance, but, in fact, he knew the words without any proper idea of their meanings. In some unaccountable way he had acquired confused ideas of them. Half an hour of patient teaching put him in possession of right apprehension of the relation of the figures to their places and meanings, after which his arithmetic went on without further trouble. An erroneous or imperfect idea of the meaning of a word or formula is often the unsuspected cause of apparently insuperable difficulty and seeming disobedience.

The future training will depend largely on the question whether the mental condition be natural or artificial, real or assumed, especially in regard to the important factor, obedience.

MORAL PRINCIPLES.—Originally acquired, like language, by the natural means of observation and inference, moral principle now becomes more definitely formulated through the medium of language, not merely, nor perhaps chiefly, by direct verbal tuition, but by inference from what is heard and read. Ideas previously acquired by means of observation and example are now either confirmed or modified, especially in regard to the supreme authority and the unseen Almighty Father from whom it emanates. The force and efficacy of the new ideas, however, are mainly dependent on the extent to which example accords or conflicts with ideas imbibed previously. The knowledge of language having originally been derived by inference from experience, this continues to be its interpreter. The meanings of words having been learned by inference from the connections in which they were used, so, in like manner, principles embodied in words receive their practical interpretation from conduct. Hence, a child's ideas of moral principles will be found to accord more nearly with their practical exemplifications than with the verbal formulæ. This remark, especially applicable to children at the age we are now considering, is not confined to them, but applies in a greater or less degree to people of all ages and religions. For instance, Christians of high intelligence, and well informed, form their ideas of the Mohammedan or Buddhist religions by their practical exemplification in the manners and habits of those who profess these religions. And these think of Christianity as they see it practised by Christians. Hence each thinks very ill of the other. The Christian missionary thinks the Buddhist an idolater, because of the image in the temple and the offerings laid on the shrine. The Buddhist retorts that the genuflexions and worshipful signs the Christian makes to the figures and emblems in the east window of his church and on the altar are idolatrous. Neither appeals to the written doctrine. Children follow the same rule, and form their ideas of moral principles as they are exhibited, rather than as they are taught.

At this age children are taken to a place of worship, and they form their ideas of worship by what they see far more than by what they hear. As sentence after sentence of a sermon follows in quick succession, the attention required to apprehend them is far beyond the capacity of a child. Indeed, very few adults really apprehend the half, and of that only a fraction is definitely formulated and fixed in the mind. The attempt to follow the sermon soon becomes wearisome to a great majority of hearers, who let their thoughts wander. Children do the same, and are often very much scolded therefor. Where the greater part of the service is formulated, as in the English Church, that part may be followed without effort, as the ideas are already fixed in the mind and apprehended, needing only to be recalled. The advantage of this formulation is particularly evident in certain situations. For instance, on a sick-bed, or abroad, a man who perhaps has thought but little of religion when in health and activity now finds the stereotyped liturgy still in memory, embodying in clear and beautiful form the vital truths of his neglected religion. And in the same manner many a youth remembers, when far from home and friends, the example of a mother in whose habitual conduct was stereotyped the vital principle of Christian faith. By that conduct her words of love and her verbal teachings are practically interpreted, and the interpretation continues to be a permanent and effective possession long after the verbal formulæ have ceased to be remembered.

HABIT OF THOUGHT.—In the gradual process of the development of the mental powers and in the accumulating of ideas, the progress is not exactly the same in all directions, but follows certain lines. These are determined partly by the nature of the environment and partly by inequalities in the faculties themselves. Internal and external causes both operate in imparting direction to these lines or inclinations. Both the nature of the subject-matter on which the faculties are exercised and the manner of dealing with it vary in every individual. Hence, certain kinds of material and certain methods of thinking prevail, and form the personal idiosyncrasy or character in each case. These lines begin to be formed with the earliest mental efforts, and by the time of which we are now treating they have necessarily become more or less fixed and decided. The growth of the powers has, in fact, been proceeding upon, and so deepening and establishing, these lines. The edifice of the Ego has been constructed upon them, and they have determined its shape. The further development will necessarily proceed on the habits already established. The form it may ultimately assume may be infinitely varied, but the type is already determined, and will not be materially altered unless by some exceptional and powerful influence, such as severe illness, accident, or nervous shock.

The physical development is also subject to similar variation during growth. Certain disproportions or tendencies to malformation, threatening weakness or disfigurement, are apt to appear in children, in which case the utmost anxiety is felt by tutors to remedy the defect, or to correct the wrong direction of the growth. The corporeal frame is always vigilantly watched and jealously cared for; but, all the time, growing defects and deformities of mind are, generally at least, neglected, and left to take care of themselves. The symmetry of the physical structure is deemed worthy of all care and vigilance. Hence, children grow up with straight bodies and crooked minds; with vigorous, well-balanced physical faculties, and with mental powers misshapen, disproportioned, and irregular. If obliquities of moral or mental habit be observed at all, they are too often regarded as accidental or inherited from some conveniently remote ancestor, and are left to be eradicated at school, where the crookednesses of the original haphazard process of mental growth are expected to be hammered out. If the pupil does not come back from school straightened and squared, the fault will probably be laid upon the imperfection of school discipline.

This being the first period in which a full and intelligent use of language has been enjoyed, and its natural consequence, inquisitiveness, prevalent, the habits formed, as well as the progress made, will have been determined by this agency. The child at the end of it will be open, candid, and communicative, or else secretive, and reticent, according to the treatment he has received and the example he has imitated.

SUMMARY.—The progress usually made during the term of about four years of childhood is both physically and mentally very strongly marked.

1. The corporeal powers have been carefully watched, exercised, and trained, and, unless impaired by sickness or accident, are agile and vigorous.

2. The ideas previously formed by observation and inference from individual experience, and confined to what could be so learned, have been expanded, by means of language, so as to comprise a great amount of knowledge from the experience of others. The infant's ideas were chiefly his own; the child's, embrace and enter into the labours of mankind.

3. Personal experience has been more influential than ever, but physical sensibilities have almost ceased to operate, and mental susceptibilities have

been modified by reason, and have resulted in enlarged ideas of the conduct and principles of others, especially of tutors.

4. Written language has been acquired. Reading can now be done with sufficient facility to be understood as well as perceived. Writing and the language of numerals, forming elementary calculations, have been added to the list of effective agencies for future progress.

5. The formulative faculty has acquired a high development by constant use, and by means of the aid and facility afforded by speech. Ideas are therefore more complete, as well as more definite and more deeply impressed. Memory has a firmer and more lasting grasp of the mental stock.

6. The Will is now definitely fixed, but whether outward conduct be regulated by it, or by the superior force of discipline, remains to be seen when the rule of tutors and governors shall have ceased to predominate.

7. The restraints of discipline, reason, and regard for consequences throw an external covering over the internal feelings and the relations of the youth with the external world, and this covering conceals the inner workings of these influential feelings. Henceforth the inner mental condition is more or less secret between their possessor and his Maker.

CHAPTER IX.

ON THE NATURAL HISTORY OF THOUGHT.

FOURTH PERIOD: YOUTH.

This period comprises school-days—the work of education in the common acceptation of the term. The youth goes to school to prepare for a future independent career of business and the duties of life. This is the first step towards that independence to which he already ardently aspires, and the hope of which constitutes the dominating influence of this period.

From the time the youth enters upon school-life he looks forward to the time when he will emerge from the restraints and discipline of tutelage and begin his own independent career. He regards the end of his school-days as the beginning of his own life. Whether he address himself to his work with a resolution to qualify himself for the business of life, or undertake it perfunctorily, to get through and to be rid of it; whether he try to do as much or scrape through with as little learning as possible, the end he has in view is his independence.

Ambition and aspirations for freedom are natural to the period and position of the youth, and are therefore to be respected. It is reasonable and desirable that he should look forward to the life for which he is now preparing. His ideas of that life which he anticipates so constantly may be vague, and are probably very different from what he will realize; nevertheless, they have a powerful influence on the direction of his thoughts and on the spirit in which he does his work. They are the result of the collective experiences of the three periods through which he has already passed. Each of these in turn since the first was based on the groundwork laid in the one preceding, and this last comprises the sum of the whole. During those years he has collected a large and varied stock of ideas of men and things, has acquired a knowledge of language in both its spoken and written forms, and is therefore furnished with the means of learning all that history, science, philosophy, and, in short, all that the accumulated experience of mankind can teach. But, along with these qualifications for the work of this critical period of his mental history, he has also imbibed other ideas of a personal nature, which will be far more influential than his stock of knowledge in giving direction to his aspirations and tone to his work. These will determine the use he will make of the means he has acquired, and the purposes to which they will be applied. They are the elements of his past experience which will now become the actuating factors of his present conduct.

OF THE CORPOREAL FACULTIES.—Learning and mental discipline being the special business of school-life, the physical conditions most favourable to such work deserve primary attention, both in regard to the bodily vigour of the pupil and the conditions of his environment. The vital importance of this factor of the educational process is now very generally recognized. In all good schools the corporeal faculties are exercised and invigorated by regular athletic sports and competitive games, which afford also good opportunities for the observation of character as well as for physical training. The practice, now general in schools, of the masters entering into, and interesting themselves in, these exercises is most salutary in its influence on both. Discipline is thus aided and tempered by sympathy. In the work proper to this period, and in the discipline necessary thereto, the relations of teacher and pupil are sometimes strained; and, seeing that the success of both depends so much on the feelings mutually entertained, it behoves masters to avail themselves of every means of attracting and maintaining the confidence and affection of their pupils. The opportunities afforded out of school are therefore of great value. Youths are particularly appreciative of interest shown in their work and proceedings, and especially of consideration for themselves personally. This, in fact, is the form in which their susceptibilities are now most manifest. Hence, confidence may thus be either won or lost by sympathy or indifference in regard to personal treatment.

The nervous system is always a highly influential factor in mental work, and one which requires consideration, as its condition varies much in different individuals even where no positive malady exists. Almost every youth has some idiosyncrasies more or less affecting the nervous system. These are often induced by shocks, frights or other undue strain of the nerves in childhood. But, howsoever arising, such nervous affections require consideration, the want of which may not only hinder work, but aggravate the malady itself.

Youths vary considerably in the relative aptitudes

of their faculties and in the avenues by which their understandings are most accessible. One, for instance, who has been accustomed to learn through the medium of the eye may seem dull in apprehending by the ear, or *vice versâ*. One assumes a particular attitude or gesture when thinking, and is put out if that be prevented. Another must have something in his fingers. Such habits may be, in certain cases, objectionable; but seeing that they are for the time influential, and the work to be done important, correction should be gradual and considerate. Mental work being paramount, it should be encouraged, and the utmost care should be observed to lighten and sweeten it. For this purpose the physical conditions, not omitting that of the ruling organ, the stomach, are worthy of special and constant attention.

OF THE WILL.—By this time discipline, social obligations, and fear of consequences have become influential in modifying the assertion of the Will, whether the true spirit of obedience be present or wanting. Nor can the actual motive for seeming obedience be now easily recognized, for that is often the readiest and most compliant which is dictated by policy. On the other hand, that may be tardiest in which moral principle has to overrule temptation. Of the two sons mentioned in Scripture, the one who acquiesced at once failed in the performance, whilst he who flatly refused at first was the one who afterwards obeyed and went to the vineyard. But, whatever may be the actuating principle concealed in the breast of the youth, the only treatment tutors can safely adopt for securing obedience is that of patient remonstrance and calm reason. Force, bad at all times, is worst of all when it can be resolutely resisted, and when it can prevail only by recourse to superior might.

The means adopted in the training of the lower animals are well worthy of attentive consideration by teachers of youth. The feats performed by elephants, lions, dogs, monkeys, and other animals are striking proofs of the influence men may acquire by suasive and gentle measures over creatures whose power of limb or physical resources would enable them to offer effective resistance. In all these cases the Will is subdued by suasive measures, commenced betimes and pursued with judicious tact. It is true that force is sometimes employed, as it was in the case of "Blackthorn" (Note 1, Appendix), but it is perilous and only partially efficacious, as was proved by the narrow escapes his owner had.

Considering what are the proper functions of the schoolmaster in imparting knowledge to qualify youth for the work of life, it is manifestly most important to both master and pupil that the path to the attainment of the requisite knowledge should be free from every unnecessary obstacle. The training of the faculties during the three preceding periods should be specially such as to prepare the way for this final work of education. Nature has adapted the physical conditions to the purpose in a manner at once obvious and admirable. The temper, disposition, and Will were first initiated whilst the conditions were suitable—that is, before the corporeal faculties were available for use. Thus a favourable state of mind was established whereby it was fitted to receive its proper furniture. Then, when the use of the faculties had been acquired, they began the work of acquiring knowledge—first by observation and inference and by such means as the child's own experience afforded. During this period speech is acquired, which initiates a new form of learning, and enlarges the sphere of knowledge beyond the range of personal observation. Now, if these several opportunities have been improved in preparation for the work of education, the master and pupil will meet on fair grounds. The one will be ready to give and the other waiting to receive the coveted treasure. If, on the other hand, the temper and disposition have been marred, and the Will neglected and allowed to predominate, then master and pupil are at once placed in false positions. The proper duty of the master has then to be deferred to other preliminary duties, for which the proper time and opportunity have passed. For one pupil who is reasonably prepared for the work of the period, two at least come without due preparation, or even with strong disqualifications. One is sent from home because he can no longer be endured there. School has been threatened so often that it has come to mean to him a place, not of learning, but of punishment, and the youth arrives there fully imbued with that idea. Another has been drenched with ineffectual lessons until the very mention of a lesson is loathsome. In one form or another *unwillingness*, the very worst obstacle to the work of the period, is set up with more or less force in many of the pupils who are sent to school to be instructed. It is not ignorance which the schoolmaster has to overcome, but unwillingness, in nearly all cases where difficulty is experienced. Nor is it a taste for learning that has to be inspired, but a rooted distaste that has to be eradicated. The chief work of the master is not in what he has to do, but in what must, if possible, be undone. The schoolmaster does not now receive the raw material on which he has to work either in its primitive, plastic state, or even with the rudiments of a correct form, but after the first processes of its formation or malformation, have already given it a rigid shape, which must embarrass and may quite frustrate his best efforts.

OF THE MENTAL FACULTIES.—The mental powers are now well developed, and the regular process of cultivating them is the special work of this period.

From the first dawn of intelligence these faculties have been exercised on two kinds of subject-matter, one personal, the other external. The former produces certain relations of the child to persons generally and individually, the latter results in the acquisition of ideas of men and things. In the former, personal feelings are the sole agents; in the latter, they are much less influential. Indeed, it may be here again remarked that the mental effort by which knowledge is first acquired is entirely free from strain, ruffled feelings, or trial of temper. So long as mental work was effected by the spontaneous act of the child himself according as it was attracted by the presentment of objects or ideas to his mind, it was free from difficulty and strain, and was at once agreeable and effective. The *grief* of learning begins with teaching. Till then, the process of learning was an agreeable mental exercise, practised by choice, and a source of almost invariable pleasure and satisfaction. Nor is it otherwise until it becomes associated with the element of personal feeling evoked in the processes of teaching. Then it too generally happens that a positive distaste for knowledge commences. Hence it follows, certainly, that during the period of school-life, when learning is the special business, success must necessarily depend to a great extent upon the state of mind induced by the kind of teaching previously practised—in other words, upon the attitude of the learner towards his instructors.

Educational processes, so far as they relate to the furnishing of the mind with information, have received so much attention under the stimulus of private enterprise and of public influence that human ingenuity can do little more to facilitate the acquisition of knowledge in the way of further appliances. Much yet remains to be done, however, in the training of the mind itself, especially in the preparatory processes of the first three of our periods. Natural appetite may thus be stimulated and fostered, and the power of mental digestion improved. Experience proves that through life from infancy the mind is naturally prone to learn in its own way, and that there is a natural desire for certain kinds of knowledge. A great proportion of all human knowledge is in fact so acquired. If men would appeal to their own personal experience, and consider the effect upon their own minds of being urged or compelled to learn what they do not want to know, they would perceive that the surest way to quench appetite and create disgust is by drenching the unwilling mind. They would then be more considerate for youth, and would apply their efforts rather to creating appetite by appropriate means at the proper time than to forcing lessons upon unwilling minds.

Susceptibilities, first physical, and then mental, but always *personal*, have been operating powerfully all through the previous periods in moulding and giving a definite shape to the character of the child. From his very birth his mental nature has been uninterruptedly affected by his personal feelings, and the result, as now developed, is the most influential factor of his present mental condition.

Personal sensibilities change in their character from stage to stage of the mental progress, but are always, in whatever form, the most potent elements in the formation of the individual edifice. They are, as a rule, from first to last, but more especially in the later periods of childhood immediately preceding schooldays, under-estimated where they are at all regarded. Many persons habitually slight children on principle, for fear, they say, of making them conceited. Thus, to avert an imaginary consequence, they impose a real evil. Conceit comes not from without; it is inspired from within, and has its history; but, if there were any real force in the excuse alleged, we think it better a youth should be a little conceited than be cowed and humiliated; better think rather too highly of himself than lack the natural principle of self-respect; better he should be on rather too good terms with himself than be on bad terms with the world generally.

Pride, self-respect, and kindred feelings, suggested by consciousness of power and by natural aspirations for independence, are very sensitive in youth, and exert a strong and generally a beneficial influence on the tone of character. These feelings are not always manifested as strongly as they are felt, because they are generally restrained by the want of sympathy on the part of tutors. The common practice of regarding children's speeches as of no account, and of silencing them with scant courtesy, puts a check on childish communicativeness. Hence youths have not uncommonly acquired a habit of reticence and learned to keep the workings of their inner thoughts to themselves. Wherever this habit prevails, however, feelings exert a more powerful influence on character. A secretive habit, the natural consequence of previous injudicious repression, is regrettable, as being opposed to that frankness and candour which are so great an ornament and attraction in personal intercourse.

The consideration for personal feelings which has been urged so strongly in preceding chapters had in view their bearing on this critical time, and still more critical process of education. The state of feeling, or rather the state of mind produced by feelings, with which youths embark on school-life goes very far towards determining the success of the work. Hence the importance of special attention throughout the preliminary stages to produce conditions favourable to the effective performance of this crowning work, and to avoid whatsoever would tend to create distaste, or oppose any other form of obstacle to learning.

There is a considerable analogy in the growth and nourishment between the body and the mind. Appetite gives relish to food, and, if this be suitable, it will be digested and assimilated naturally. If, on the other hand, there be no appetite, or if the food be distasteful, relish is impossible and disgust inevitable. Neither digestion nor assimilation is natural under such conditions. Even the best of food may be made loathsome by the manner of administering it; as when a strong-handed disciplinarian ordains that a certain portion shall be eaten *nolens volens*. If not eaten at breakfast, it shall be served at dinner, and no other given till it is consumed. It is just so with mental food. Naturally attractive, it may be so administered as to inspire disgust. The inexorable discipline which despises appetite and enforces rule against Nature will convert what might be agreeable lessons into intolerable tasks.

External objects become invested with increasing interest from year to year in proportion as knowledge and the uses to which it may be applied increase. In the first two periods, observation of whatsoever presented itself and captivated the attention was busily employed, and new ideas were accumulated rapidly; but, in the inexperienced condition of the faculties at that time, ideas were crude and imperfect, being formed on external aspects only. In the third period, they were corrected and qualified by means of enlarged experience, the use of language, and the exercise of reflection. Still, mental effort hitherto was almost entirely spontaneous, being attracted by objects or ideas incidentally presented to the mind. The first attempts to *exact* attention to particular subjects introduce a new form of mental exercise. Till then, learning had come as it were spontaneously; but with the first lessons learning wears a new aspect, as requiring *attention*.

The effective formulation of ideas exacts either fixed attention or frequent repetition of the presentment. The ideas of the early periods are weak for that reason, and would be of little avail without repetition. Those of a very striking or attractive character are exceptional, and are deeply fixed because of the attention they excited. Hence, for school-work attention is very necessary, and a good deal of repetition too. The work of this period is to classify, correct, and complete the ideas previously and imperfectly formulated, and to define and fix them in the mind in such manner that they may be available for use in after-life, and also to acquire such other knowledge as may be necessary in each particular case. For this purpose it is not sufficient to follow the former easy method of procedure. Attention is no longer free to take what is most attractive of things incidentally presented to the mind, but must now be under control. The power is now required of diverting the attention from all other matters, however inviting, and of directing it to a given purpose. The mental effort involved in this act is very considerable. It is a rare accomplishment, even in adults, to concentrate their attention on any given current of Thought for a length of time. To youths, especially to those who have not been accustomed to such effort and to those who have a nervous temperament, it is a severe strain, for which due allowance should be made and every help and encouragement given. The difficulty of giving sustained attention is generally under-estimated, and teachers are apt to regard failure as indicative of carelessness, incapacity, or perverseness, when the defect is not in the Will, but in the power of control.

Another very essential condition of success in the work of education is, that youths should be impressed with the nature and purpose of the work they are required to do. The same intelligence which qualifies a youth to receive instruction would enable him to understand and appreciate the use of what he is set to learn if it were properly explained to him. In our experience we have found few cases in which any but advanced pupils had a proper idea of the practical use of the things they were required to learn. Hence, to many pupils the dry rules of grammar and their lessons in the dead languages are regarded much in the same way as shot drill is regarded by convicts—as useless and humiliating drudgery. To a reasonable being a reason should surely be apparent for any effort exacted of him, and, in the absence of such reason, he could scarcely be blamed for the most perfunctory performance. Earnestness of purpose cannot be expected where the purpose itself is vague, shadowy, and doubtful. Yet in all the affairs of life there is not one for which earnestness of purpose is more essential to success than for the effective formulation of ideas. Of all the expedients by which a youth may be best induced to give his heart to his work, none is more effective than to convince him of its practical value. The work of committing to memory the dry rules of grammar, especially those of a dead language, and the various methods of manipulating figures, are not attractive work, nor such as could easily be made palatable, but it may be relieved of positive irksomeness by having a definite purpose in the mind of the learner. This is exemplified in the case cited in chapter xv.

It is desirable that the pursuit or profession of the pupil should be fixed as soon as may be, in order that education may be adapted accordingly. The choice, too, where practicable, should be decided by the pupil, who, in that case, should be made acquainted as thoroughly as possible with the circumstances affecting it, in order that the choice may be determined on reasonable, not on fanciful or mistaken grounds.

MEMORY.—The stock of ideas acquired during previous periods, and now available for the work of education, consists almost entirely of such as are under the ready control of memory. Those which are not so at command are of little, if any, practical value. The nature and extent of the ideas so in possession determine the qualification of the youth for the work now before him. On this depend the receptivity, intelligence, and fitness of his mind to profit by the instruction he is now intended to receive. If it were possible to take the stock of his available ideas, and to arrange them in the order of the prominence of the place they take in his mind in such manner that the nature and relative force of each could be seen, the result would show exactly how far previous training had served its purpose. It would then be apparent whether the equipment of the youth for the purpose of this culminating work was well or ill adapted to its end. As no such analysis is possible, however, either for a particular case or for a general average, we must be content with such an approximation as inference from experience may furnish.

The stock, we know, will in every case consist of ideas of two kinds—that is, of personal feelings and of external persons and things. Further, we know that the former will be represented by temper, disposition, faith in, and obedience to, authority; and the latter by items of general knowledge, of which language will be the chief. All the stock of available Thought will be comprised under these heads in every case, but the relative force of each will, of course, differ widely, according to the circumstances in which the ideas themselves were formulated.

Of the two principal classes of Thought there can be no doubt as to which will be most influential both in deciding the character and suggesting the proceedings of the youth. Hence, now that he is brought face to face with the first real work of his life, it is manifest to which of two classes of Thought the previous training should most properly have been directed.

As regards general knowledge, the present state of the youth's stock in command will, on analysis, show that those ideas which are recalled most easily, and which in fact are always ready, are such as are in constant use, and therefore could not be forgotten. Of these, language is the chief, and the familiar objects, persons, and ideas of every-day life are nearly as ready for recall. These, therefore, afford no test of particular power of memory, for they are the common property of every one, including persons of weak intellect. A special value of language in relation to power of memory consists in the evidence it affords of the power possessed in infancy. For, as the acquisition of language exacts a very great exertion memory in the first instance, and could not possibly of be acquired otherwise, it is thus proved that, during that time at least, this faculty of the mind was vigorous. Any subsequent failure, if such should occur, must therefore be referable to special causes. These are discussed in a succeeding chapter. The point pertinent to our present purpose is, that under certain conditions a great work of memory has actually been performed in every case where language has been acquired, and that, therefore, the power of the faculty has been clearly proved. Whatever those conditions were, they have been proved effective, and should as far as possible be reproduced for the purpose of fixing permanently in the pupil's mind all such further knowledge as is to be imparted. General rules only are discussed in mnemonic systems and in our chapter on Memory; but particular points are of importance to individuals, as being the result of personal idiosyncrasy. The special circumstances of each case are therefore to be considered in applying general rules to such case.

Reflective faculties being exerted on the stock of ideas in possession, the range of their operation extends in proportion as this becomes enlarged, and its nature depends on the kind of ideas that are most deeply impressed and most readily remembered. The currents of Thought which predominate in the mind, and the habit of thinking, will determine the general tendency of reflective action, and will go far to decide its force and effect. Hence it follows that the ideas in present possession at any time influence by reflective action all those subsequently acquired. The ideas formed during each stage, from babyhood onwards, affect those of succeeding periods. Hence the reflections of youth derive their character from the ideas acquired previously, and more particularly from those most recent and strongest ones of the period immediately antecedent. Throughout the mental history, each stage depends very much on those preceding both for the progress made and for the direction that progress takes. In short, character is built up on the lines first laid down, with such modifications only as may be produced by forces of such exceptional strength as to overcome the naturally acquired bent.

Whatever may have been his antecedents, the youth's most influential ideas will almost always be those of personal feeling. The state of his mind as regards external persons and things will be mainly due to the personal feelings induced by his previous experience, even in those cases where a strong and decided taste for any particular pursuit may have been established.

Now, the tendency of personal feeling, as a motive or prevailing influence in reflection, is to induce an unwholesome and unprofitable kind of Thought which substitutes, for healthy, active, reasoning exercise, a ruminating form of reflection—a brooding over, and indulgence of, whatsoever feeling may be most excited. Seeing that reflective processes are the means by which observation is most effectively aided, supplemented, and turned to account, and whereby ideas are corrected, expanded, and completed, it is most desirable that they should have a wholesome direction, and should not be diverted into a channel from which nothing useful or beneficial can be hoped. The best means of inducing a salutary action of the reflective faculties is to keep them actively employed in whatever form of useful exercise is most attractive in each particular case. Education has for its objects—(1) the acquisition of such knowledge as is necessary for the conduct of life; (2) to establish such discipline of the faculties as will strengthen them and direct their application according to sound principles, both mental and moral. Hence, a good deal of the tuition imparted for this latter purpose has little value apart from the discipline it promotes. Such being the case, the nature of the particular study or pursuit is of less consequence than its efficacy as a means of mental culture. A youth had better, in fact, be engaged in almost any healthy pursuit than be suffered to lapse into unwholesome broodings. The reflective faculties are those, therefore, which need, in this stage of mental progress, the most assiduous attention and culture.

In the exercise of their highest functions, their proper purpose, the reflective faculties afford the means of discerning truth and detecting falsehood, of discriminating between good and evil, and of extending knowledge, by processes of reasoning, beyond the superficial qualities of things to the hidden forces and laws which regulate their mutual relations. Art, science, philosophy, and religion owe their progress and development to the power of reflective processes. As the laws of gravitation were discovered by the series of reflections suggested to Newton by the fall of an apple, so the higher nature in man owes its conquests to the use of these noble faculties, and the lower nature achieves its predominance by means of their abuse.

Judgment is the weak element of juvenile character, the faculty which most needs the ripening effect of wide experience. Nevertheless, opinions, inferences, and judgments are very freely formed, and confidently indulged, at the age we are now considering. Their influence on conduct is also very considerable, as it is proportionate to the independent feeling that is now predominant.

Judgments and opinions formed in youth are generally more decided and confident than those of later life, because they are formed on fewer and simpler data. It is far easier, in fact, to weigh a few items of evidence than to deal with a large number. Hence, inference is easy and correspondingly clear and certain to the mind of the judge who has but few data to consider. The difficulty arises when, in after-life, accumulated experiences furnish a large amount of evidence of a conflicting nature. The less he is embarrassed by evidence, the more confident is the judge, and the more obstinately does he adhere to his verdict. Opinions and judgments are, therefore, very influential factors in the proceedings and progress of youth during school-days, and need to be treated judiciously. They should be rather corrected by convincing evidence than contemned and contradicted. As character is now in advanced embryo, its further and final development will be determined mainly by the power of this faculty, which, therefore, is of paramount importance. In the exercise of this faculty tutors may see the most influential results of previous training or want of training, and may thus see the particular defects which require to be supplied, the forms of error to be corrected, and the direction in which cultivation was most needed.

TEMPER is a very formidable factor in the work of school-days, and is, in various ways, a great help or a serious hindrance to success. It is subject to restraint and modification by the joint influences of example, precept, and the obligations of social discipline; but, though these influences may serve to prevent outward manifestations, the inner feelings will still be those originally impressed and since established. However it may be controlled or suppressed by modifying circumstances, its original character will be maintained. Still, it is even yet more amenable to correction than it will ever be at any future time, and the most powerful influence that can be employed in its subordination is example. The exhibition of undisturbed patience by tutors under severe trials, such as are not infrequent in school-work, is often very impressive, especially upon on-lookers. As an example of forbearance and consideration exhibited to a member of a class, its influence on the witnesses is most beneficial, and, though it would not eradicate bad temper in any of them, it would suggest effective control.

Faith.—The individual influence of tutors is still mainly dependent upon and proportionate to the confidence they inspire; but, as the power of reason increases, and experience becomes enlarged, faith begins to be divided between persons and principles. In proportion as sense of duty, the obligations of moral principle, and the weight of friendly counsel prevail over personal feelings and influences, the mind naturally seeks some standards less fluctuating, more definite, and more authoritative than the imperfect models that have served as the ideals hitherto followed. Faith loses none of its influence; on the contrary, it has ever-increasing force, but is fixed on different standards. The influence of tutors is no longer dependent solely on the faith reposed in them personally, but on the principles they represent and the conformity of their conduct to those principles. The growth of belief is a slow and probably, in some cases, an unconscious process, the product of experience, and is not demonstrative. The actual condition may be—indeed, generally is—unobserved, for the opinions and ideas of youths are generally held of little account. The time we are now considering is very often a turning-point or crisis in regard to the vital matter of religious belief. If faith in God be not now recognized as the foundation of all authority, faith itself is not therefore extinct, but will be fixed on some other ideal. It may be transferred from one ideal to another, and be more or less unsettled, but it is not obliterated. It is a natural principle, which, if not fixed effectively on the Supreme Being, tends to idolatry in some one or another of its many forms. If faith in God be shaken by partial views of His providential dispensations, or by doubts of His justice and love as therein manifest, it may, and often does, rest in the outward formularies and observances of religion—a veritable development of the old idolatrous tendency—or it may take refuge in the laws and forces of Nature, which seduce many adult waverers. Once unfixed, it may seek refuge in one figment or another, and find a final rest in *self*—the last resort of the destitute in faith. We have in our minds a youth of high aspirations, who has exhausted all other modern resources as an anxious inquirer, and is now reduced to this poor extremity. He believes in nothing outside of his own being. The condition of a youth who begins his independent career of the business of life with faith in nothing better than himself shows the importance of early cultivation of the vital principle of faith in God. Whether his mind have been unsettled by observing the hollowness or inconsistency of profession or by an example of indifference and practical unbelief in God, by reflections of his own or suggestions from without, his state is one to excite the deepest sympathy; and it is unhappily not now a case of uncommon occurrence.

OBEDIENCE.—A genuine respect for authority and a spontaneous desire to please, which constitute the spirit of true obedience, are high qualifications for school-work. In the absence of the real principle, an outward conformity to rule is generally substituted. Openly rebellious and refractory spirits are too intolerable to be permitted, and are therefore reduced to seeming submission by coercive measures. In general, however, outward conformity to discipline is assumed, and often simulates the genuine principle so well as to deceive tutors and others. Apparently sincere, because of its having become habitual, the real character of such conformity is revealed only when relaxation of restraint or strength of temptation exceeds the force of habit. There is generally a ready obedience available for school-work, but the manner in which this work is done depends in a very great degree on the animating spirit, whether it be on principle or by the acquiescence of expediency. The mental progress will be satisfactory or otherwise according as the work is done with a will or merely to save appearances and escape disgrace.

Hence, also, external regard for the observance of religious exercises is sometimes so well and consistently maintained as to establish a character for sincerity which, after all, proves under trial to have been superficial and wanting in the true principle of obedience to Divine authority.

MORAL PRINCIPLES.—Originally derived from observation, the ideas of moral principle have been, during the previous mental history, gleaned from a mass of experience. The greater part of this observation was made before language was sufficiently understood to be of any avail for preceptive teaching.

Thus moral principles, like words, have derived their significance, and their terminology has received its interpretation, from the experience which, during these long years, has been always exhibiting them in their most practical form. Now, however, that precept is understood and judgment is freely exercised, the youth forms conclusions of his own, and reduces to general principles the exemplifications of his past observation. In the exercise of his independence these conclusions will go far to determine his conduct, but whilst he is under discipline this may still be conformed thereto. The inner workings of Thought, the secrets of the heart, will be in or out of harmony with the outer conduct according as the observance of rule has been dictated by principle or expediency.

The quality and force of moral principles imbibed during previous years are first tested practically when the direct influence of tutors and the restraints of discipline are removed. In the meantime, therefore, their supremacy will be determined or their subordination to expediency will be established almost entirely by the experience and intercourse of the first periods.

HABIT OF THOUGHT is an exponent of character. Nearly every faculty, mental and physical, is by this time moulded more or less definitely into the form of habit. The manner in which almost every act is done, the mode of proceeding by thought, word, or deed, is dependent to a greater or less degree on habitual practice. Life has already begun to run its course in lines that were first imperceptibly traced, and have since been deepened into grooves. In so far as these may give a salutary direction to Thought, word, and deed, mental training will have been useful and successful, and *vice versâ*. The training, whether it have been by judicious care or by culpable neglect, is now crystallized into definite shape and exhibited in character. Even the Will, the master-power of the mental estate, runs with smooth, almost mechanical motion under the influence of habit if faith in, and obedience to, authority have been happily established during that appointed time when, in the order of Providence, the natural conditions were so favourable for the purpose.

SUMMARY.—It was comparatively easy to summarize the mental progress of the first periods of our history, but the difficulty increases with each successive stage, and it has now become impossible to state with even approximate correctness what the mental condition may have become at the close of the educational course. The different kinds and degrees of mental attainment acquired during this period depend upon so many diverse factors that mental condition could not be reduced to any general formula. The kind of knowledge acquired will have been decided by the social grade in each case and by the future avocation each is destined to follow. But whatever the social grade of the youth, his natural capacity and the kind of pursuit for which his education is intended to qualify him, the success of his instructors and the eventual mental attainments of the pupil, will be determined in a far greater degree by the disposition and qualifications with which he entered upon his school-life than by any efforts his instructors could possibly exert.

Whatever may have been his natural mental calibre, a good temper, docile disposition, and attentive habit would facilitate the directing of the youth's faculties, further the aims pursued, and turn to account his powers of mind; whilst, on the other hand, the alternative disqualifications would embarrass the work of the period and effectually hinder the progress of mental attainment.

At the close of the educational course the mental condition will be indicated in three principal ways— (1) by the kind and extent of the knowledge acquired; (2) by the discipline of the mental faculties, and consequent power of employing such knowledge; and (3) by the nature and force of the moral principles imbibed. Of these, the last-mentioned is that which will be most influential in deciding the future character and conduct, and in giving direction and purpose to future effort. However important the first-mentioned may be as a qualification for future avocations, its influence on mental progress will depend mainly upon the discipline of the mind and the acquired habits as regards control and attention. Hence, the future development of the inner self, after the youth shall have attained the independence of action for which he has sighed so long, will depend less on the store of information he may have acquired than on the discipline of his mind and the moral principles by which his efforts will be directed.

Such being the relative importance of these three factors of the individual estate, such has been the order in which they have been regarded in the suggestions we have made in these pages for the preparatory processes of mental culture. Such, however, is not the order of the estimation in which they are generally held. Practically, this order is inverted. The first place is usually accorded to the attainment of general information. Parents generally so measure the value of the work done at school. In like manner the qualification of candidates for the public services is similarly tested. To us it seems strange and inconsistent that so much reliance should be placed upon a qualification which is itself generally of so fugitive a nature. Still more so is it that the more permanent and more important qualification of mental discipline should be so lightly esteemed as to be virtually ignored.

The estimation in which youths themselves hold the qualification to which such paramount importance is attached by their parents and public employers may be fairly measured by the way in which it is generally regarded after it has served its purpose of delivering them from tutelage or securing their admission to some public appointment. There are probably few who go so far as the youth mentioned in chapter xiii., who buried his books irrecoverably in a deep hole; but a very great many who do not perform that ceremony treat the lessons they have learned so laboriously at school or college with equal neglect. The proportion of those who afterwards pursue the studies they began at school to those who forget totally all but what their daily duties keep bright by use is extremely, significantly small.

With regard to the third-mentioned of the foregoing factors, the one which involves the actuating principle of the individual estate, and which for that reason we hold to be of paramount importance, little can be positively known in this stage of the mental history. Except in regard to appearances, it has by this time generally become a close secret between the youth and his Maker. The anxiety of parents in this respect, however, is not as a rule so much exercised in regard to the general principles of good and evil, as these are set forth in the Christian faith, as it is for the maintenance of the particular sectarian distinctions to which they give their own adhesion. The natural effect upon the mind of youth of thus setting a practically higher value on the mint and anise than upon the vital principles of judgment and faith is a terrible disparagement of these weightier matters, and tends to lower incalculably the true estimate of religious truth.

CHAPTER X.

REVIEW OF THE FOUR PERIODS.

A REVIEW of the brief sketch contained in the four preceding chapters shows that the human mind is developed according to the same order which prevails in the world of Nature generally. Beginning its existence from zero, at the birth of the babe, it attains its maturity by a slow and regular process. It further appears that mental power has a certain natural relation to other forces, by which relation its place and functions in the order of Nature are determined. In the human subject the immediate relation of the mental power is with a highly developed organism, endowed with most susceptible corporeal sensibilities. These are in a high state of development when the mental force originates in their midst. The process of its growth has a certain natural relation to those corporeal sensibilities and powers. Its development is not casual, but follows a regular, natural order determined by that relation, and proceeds in accordance with laws analogous to those which regulate other natural developments in general.

In order to understand the actual place of the mental force in the order of Nature, and to see its particular relation to the human individual, it is necessary to trace it to its origin in the lower animals, where, in the order of Nature, it first makes its appearance. There its origin is very obscure, being first manifest in creatures of a very low order of organism, endowed with sensibilities of a very elementary kind and with volition of a very circumscribed extent.

In them it exercises very limited functions, being restricted to an environment proportioned to the powers and sensibilities of the organisms, and therefore attaining a low degree of development corresponding to the sphere of its operations. Thence, proceeding through many successive orders of animals of gradually increasing organism and of higher corporeal sensibilities, the mental force and functions progress in proportion to, and in dependence upon, the development of the organisms and their enlarged environment. They attain their highest power and widest sphere of operation where this is most advanced and is possessed of the greatest sensibilities. Hence it appears that the mental powers depend for their development upon the means and appliances supplied by the organism for their connection and communication with the outer world.

Man, possessing the highest development of corporeal organism and power of sensibility, is thereby endowed with the conditions by means of which mental force and functions may be exerted upon the world of Nature in its widest sense, and may therefore achieve their utmost attainments. Still, it is not by his mental powers that man is distinguished from the lower orders of intelligent beings. In these powers he excels the brutes only in degree, and but for the means conferred upon him by language the comparison even in this respect would be very different from what it now appears. The characteristic which distinguishes man from the lower animals is not one of degree, but of kind. It is not in what may be attained by the mental powers, but by the discernment of good and evil, the basis of *moral* force. It is not in a further development of the lower nature, but in the introduction of a higher, which is destined to subject and supersede that lower nature. It consists in a *new* element—*moral* force, which was introduced into the world when man first attained his likeness to the Divine nature by his knowledge of good and evil, and thus became endowed with a *conscience*.

This new factor came into operation in its appointed time and place in the order of Nature, as those other forces which preceded and prepared the way for it had also come—that is, when the conditions necessary for the exercise of its functions had been established. It appeared according to a prescribed order of the universe, as this has been displayed in the developments of material Nature by the forces successively brought into operation therein. Material order first rose out of chaos by means of the gradual operation of forces proper to that purpose. Thus through successive stages of progress inorganic Nature received such development as adapted it to vital action. Vital force then supervened, and organisms appeared upon the scene responsive to the new impulse, and vital action began its course. Thence, through progressive steps, organisms attained a condition of advancement such as to qualify them for the further forces of sensation and volition. Thus, animals endowed with voluntary activities and certain sensibilities took their appointed place, fulfilled their part, and accomplished the ends of Providence in preparing the way for the mental force. With each new factor comes a new development in which all others combine, and by their united offices and operations a new and higher order is attained. Hence the mental force prepares the way for the introduction of the moral, and this in turn opens the gate to that spiritual state in which God, the supreme Centre and Author of all the forces of the universe, sits enthroned.

Each successive step in the mighty progress depends, both for the forces which operate and for the sphere of their operation, on that preceding, and the whole form one connected, continuous chain from the zero of a material chaos to the highest forms of being known or conceivable. And as in the whole jointly, so also in each severally, the progress from its zero up to its appointed destiny is effected by means of forces operating according to regular laws, through successive stages, which are mutually interdependent. It follows that the functions of the mind are determined by the place the mental force occupies amongst the other forces of Nature, and by its relation to those forces.

Further, the growth of mental power from zero to maturity proceeds in the same manner and according to similar laws in the individual as in the natural world in general. In its earliest stages the mental power in the human subject resembles that of the more intelligent races of the brute creation. It receives the first material on which it is exercised through similar organs of sense. The first impressions made on those organs are received, noted, and formulated into ideas in the same way and by similar means. In their first mental efforts, and in the first use of their faculties, the two races run closely parallel, notwithstanding their wide departure thereafter and the immeasurable difference of their ultimate conditions—the one being in permanent bondage to its lower nature, and subject to the dissolution which awaits the material world; the other, invested with powers superior to its corporeal frame and to its mental nature, and susceptible of a higher and eternal life. The incipient mental faculty is stimulated to its first perceptions and is entirely dominated by bodily feelings and appetites, which supply the first ideas and experiences to both the lower and higher types of intelligent beings. First in point of order, the animal or sensuous element furnishes to both the knowledge necessary for their respective careers—the one to pursue and cultivate his own proper lower nature, the other to subdue and subordinate that nature by virtue of his own proper and higher nature.

The Archbishop of York, in his "Laws of Thought," makes reason the distinguishing characteristic between the human and the brute natures. This seems irreconcilable, however, with the evidences of reasoning power which abound in the lower animals. Such instances as those given in the Appendix to this work, Notes 8 to 16, appear to afford irresistible proofs of reasoning, and that of no mean order. These cases, it will be observed, are independent of tuition of any kind—the spontaneous exercise of reason without the aid of example or any form of suggestion. We do not cite in this connection any cases where tuition or training has been employed, as these, wonderful as they are, might be said to involve a transfer of the trainer's reason in some degree to the animal, and not to be of his own natural power. These latter cases are chiefly useful as proofs of the efficacy of the methods employed in teaching. The others seem to show that the mental element in the two races differs only in degree, and that the true distinction lies, as before stated, in the knowledge of good and evil, the basis of the moral force in Nature.

The knowledge of good and evil is essential to the effective choice of the former as the natural counter-agent of evil, and is the means of subordinating the lower nature to the purposes of the higher. The appetites and passions of the lower nature in man are necessary to the development of the higher, for they constitute the kingdom to be subdued and governed—the sphere in which the higher power is to be exerted. A right understanding of the Divine purpose in ordaining the higher nature and its functions requires a due recognition of the relation it bears to the lower in the order of Nature. Hence the natural history of the thinking faculty in man needs to be studied from its first dawn, under the dominating influence of the lower nature, up to its maturity, where it is invested with its higher powers. Until the birth and growth of the faculty and its gradual development are fully traced it is not possible to establish mental science on the sure basis, beginning at the beginning, and following up the process of cultivation by means adapted to the end.

Such was the purpose in view in the preceding chapters, and, whilst sensible of much imperfection in the brief sketch therein contained, it nevertheless discloses, in part at least, the secret workings of the mind which lie hidden in the obscurity of infant life. It may therefore afford some help towards a more perfect solution of the problem on which depends the only prospect of success in giving a right direction, in fostering a vigorous growth, and forming a healthy constitution to the thinking faculty from the beginning of its operation.

The successive stages of the mental progress, viewed with reference to the relation of the mental to the other natural forces in operation and to the end to be attained, will appear, by a review of the preceding chapters, to afford the most suitable means, opportunities, and conditions for the work to be done. The long period of helpless dependence and pressing needs of the babe make his condition the most favourable possible for enkindling those feelings of love, faith, and spontaneous submission to authority on which future training of the mental faculty so much depends. Again, the tentative efforts of all kinds which excite the first activities of the growing faculties are precisely such as are best adapted to cultivate and strengthen those powers and at the same time to gain knowledge of the kind most appropriate to the situation. The ideas gleaned, during the unobserved action of the mental powers, from the conduct of persons, and from the aspects of external objects of the environment, are formed under conditions so completely under the control of tutors as to afford them the means of acquiring unbounded influence over the infant.

The acquisition of oral language, the most necessary kind of knowledge for the introduction to learning and to life, is accomplished effectively by the infant mind in a manner which proves that learning is not of itself a great effort or a painful task, but a natural operation, a matter of choice, and that, when pursued under suitable and natural conditions, it is a source of personal gratification. Such, in fact, is its character in the lower animals, whose scions pick up all their knowledge (a considerable sum) without the pains and penalties so often accompanying the efforts of human tutors. Men who undertake the teaching of the lower animals recognize this as a fundamental principle of their art. The secret of their success is in the exercise of consummate patience and attractive devices. They know that their pupils may nearly always be allured to learn; but few, even of those it would be safe to chastise, could be driven to learn and practise the tricks and performances that amuse the public.

Learning and imitation are, in fact, natural propensities in both the lower and higher races of intelligent beings. This is shown as plainly by the way in which language is acquired by children as it is by the way a kitten learns to catch mice. The exercise of observation and inference by which infants acquire language is conclusively attested by their use of speech, and it is equally certain that in the same unobserved way they have also obtained other kinds of knowledge of a less demonstrative nature. It is not possible that they should have *selected* one particular subject of observation to the exclusion of all others equally accessible to their faculties. On the contrary, it is certain that, in the same unobserved manner, they also acquire a considerable stock of ideas of other kinds. Hence many of their precocities and peculiarities are attributed to inheritance, as being otherwise unaccountable.

The fact proved by the history of the early periods of mental progress—namely, that there are appropriate times for the several acquirements comprised in effective training of the intellect—is not sufficiently, if at all, recognized in the current practice, and the consequence is that the work of training the mind is generally neither well timed nor rightly directed.

In tracing the current of Thought in the progress of mental action through the successive periods of our history, only those conditions are discussed which are common to children in general. There are other conditions, however, of a more particular kind affecting certain classes of children which require notice. We do not refer to personal idiosyncrasies, which are innumerable, but to such conditions as affect considerable numbers. Of these, sex, temperament, social position, environment, and physical infirmity may be specified. The great difference of natural capacity in individuals is discussed in a separate chapter on Inheritance in the sequel. The others will be now briefly noticed.

SEX.—The physical conditions and functions of the sexes respectively affect the mental faculties as regards both their force and operation. In the lower animals, where no artificial conditions exist to interfere with the natural order, there is great diversity of physical condition and relation between the sexes. Beauty of form, colour, and adornment are as often bestowed on the male as on the female. Song in birds is most commonly characteristic of the male sex. The male spider of some species is insignificant in all respects as compared with his Amazonian spouse, who disposes of him summarily when his services are no longer required. The queen bee is altogether phenomenal, as is also the female termes, in her physical proportions and functions. It does not appear that any law can be deduced from the lower grades of animal nature to determine the mental relation of the sexes in general. The mental condition seems to depend upon, and be determined by, the physical functions. In the order of evolution the mental has been adapted to the physical conditions in each particular order. In the human races the relative position of the female is generally inferior both in physical force and mental status. In aboriginal tribes her position varies considerably. In some, she is little better than a slave and drudge; in others, she approaches equality with the male sex. In the civilized races the mental distinction is rather in the power than the quality of mental operation. Women generally excel in that kind of discernment which is independent of the reasoning powers, and is strongly marked in young children and in the lower animals. A woman sees the form and force of a reason; a man sees its anatomy. This latter is a necessary qualification for scientific research and exact reasoning, but in the ordinary affairs of every-day life it is not required. The acute discernment of women in such matters, and their independence of the anatomy of the reasons, are an advantage rather than a defect.

The training of the mental powers, with which we are chiefly concerned, is not materially affected by either the physical or functional differences of the sexes. Their education differs chiefly in the particular kind of attainment most useful in their respective spheres, and is not dictated by any such infirmity of the one or superiority of the other as would necessitate a difference.

Modern ideas tend to disturb the old order of things by conferring on women the qualifications for other pursuits than those of domestic duty, and giving them an independent mental and political status. Women are proved to be quite equal to such a position, but whether it is appropriate to the sex is another matter, and one which does not concern our present discussion. The training of female children is effected with less friction than that of the other sex, owing to their gentler disposition and more acute discernment. This characteristic is evident in their quicker acquirement of language, the great agent of mental progress. Their milder temper and disposition are due to the gentler and more considerate treatment accorded to the sex in the earliest stages of infancy. In the lowest orders, where such consideration is less observed, they do not exhibit the characteristic gentleness of the sex in the same degree.

TEMPERAMENT.—This element of individuality is very influential in mental operations. As an important physical condition it affects the nature and degree of mental susceptibility. The results of treatment depend very much on the temperament. Irritation, for example, excites a nervous temperament more acutely than a phlegmatic one. In the former, it operates quickly and demonstratively; in the latter, it produces sullenness. In one, the effect is impetuous; in the other, deep and lasting.

The adaptation of mental training to individuals of different temperament is more in the manner than in the matter of the work. The moral element is the same in all, though differing in the mode of its manifestation. Good and evil retain the same relation in all. The quick impetuosity of the one is more demonstrative, but not morally different from the more hidden and deeply seated feeling of another. In all alike, ideas of moral principles are originally derived from example, and will accord therewith, though the forms in which they are exhibited in conduct may differ.

SOCIAL GRADE.—The general influence of social position on the currents of Thought has been touched upon in a former chapter, where it is shown that though grade determines the external subject-matter to a considerable extent, yet the circumstances that influence character and conduct do not vary materially in different grades.

The tendency of modern legislation for the education of the lower classes is to assimilate the instruction given in Board schools to that imparted in private institutions to youths of the higher classes. But, whatever kind of education may be deemed necessary for youths generally, the qualification required to enable them to avail themselves of it effectively must be acquired in former stages. Under the prevailing practice, the proper conditions of receptivity are hardly ever present in children of any grade, and other conditions adverse to the processes of education are frequently established instead. The preparatory qualifications seem to be as compatible with the circumstances of the lower as with those of the higher classes, excepting only those of the very lowest and utterly degraded class.

The recognition of the appropriate times and opportunities which in the order of Providence have been adapted for the different branches of mental culture is the first step towards a reform of existing practice, and this is equally practicable and important to all classes.

Children of the lower orders have an advantage in receiving their first training from their own parents, whereas the children of the upper classes are generally consigned for their early training to the care of dependents, who neither feel nor could be inspired with the same sense of responsibility as parents. In all grades the ideas on which actuating principles depend are suggested by example, and are inferred in the same manner as the meanings of words from observation. They are, in the main, of the same character, so far as regards their intrinsic moral value, in all classes.

Immoral habits are more flagrant in the lower than in the higher classes, but this is a difference of appearance only, and not of moral turpitude. The vices of the rich are not less offensive to virtue, nor in their naked reality are they less revolting, than those of the poor. The drunkard reeling out of the ginshop and uttering his coarse language is not more repulsive than the wealthy sot whose vice is concealed from open scandal. Literary or scientific attainments and refined language do not in any way mitigate culpability nor qualify the comparison of the offence of the rich with that of his neighbour in the street. In all grades the appetites, passions, and forces of the lower nature suggest the same temptations under different forms, and, in all, the conflict with the higher proceeds from the same sources and is waged with like severity. Scripture teaches that triumph is more frequent amongst the poor than amongst the rich. The noblest Exemplar of virtue and Teacher of truth who ever trod this earth chose, in His Divine wisdom, for the accomplishment of His great work, the rank of the poorer class of mankind.

The opportunities for mental culture differ in kind in the different grades of society, but, in all, the higher nature has the same opportunity and the same difficulty in asserting its supremacy over the lower. It is worthy of note that one of the most striking and exemplary instances of true generosity recorded in Scripture was exhibited by one of the poorest and humblest grade—the widow who gave her two mites.

ENVIRONMENT, EXAMPLE, &c.—The several influences affecting mental progress respectively produced by environment, example, and the elements of evil and error, which are universally and inseparably mixed up with these factors, are not distinguishable one from another. The influences which determine the nature and degree of mental progress, character, and moral principle in each individual vary according to the share contributed by these factors jointly and severally.

Seeing that the ideas first formed in the infant mind are acquired by observation of what is within the range of its perceptions, and that the mental powers must necessarily—unless actually dormant —be exercised thereupon, the importance of the subject-matter so submitted for his contemplation, whether in the form of word, conduct, treatment, or external environment, cannot be overrated. Seeing also that a knowledge of language is acquired during the very earliest stages of infancy, the powers of observation and inference necessary for that purpose are clearly proved. It stands to reason, therefore, that those same powers were not exclusively confined to that one kind of exercise. Words were not the only subject-matter of all that came within the range of perception to attract notice or to be selected for special contemplation. On the contrary, it is equally certain that conduct and all other subject-matter of the child's environment were regarded with some degree of attention and were productive of effective results. Although other forms of knowledge may not be so easily applied nor so demonstratively exhibited, the possession of them in some degree is equally certain. In fact, the little precocities and surprises of continual occurrence, and which often seem so unaccountable, are thus easily accounted for, as are also the developments of temper and other forms of character the origin of which seems so obscure that they are almost always attributed to inheritance—that most convenient of all excuses for the results of mismanagement of children.

Hence, the influences of example, conduct, and of all the elements of environment operate concurrently and with the same force along with those from which the knowledge of language was derived. The fact that, during the waking hours of babyhood and infancy, the mental powers are thus proved to be always occupied in observation of what is within range of perception—that the elements of environment determine the subject-matter of contemplation—that, in short, before precept could reach his intellect, example and conduct have been always instructing, and the young mind has always been imbibing the practical lessons so imparted—is the most impressive of all the facts of infant training and of tutors' responsibility. Add to this the further fact of the universal and inevitable admixture of evil and error, of what is wrong in principle and delusive in appearance, in all the elements of environment relating to both persons and things, and it will suggest most serious considerations respecting the conditions imposed, the functions fulfilled, and the influence exerted by environment, in the comprehensive signification of the term, on the nature and degree of mental progress in infancy. The responsibility of tutors during the infant's helplessness and dependence manifestly includes the important duty of adapting the environment to the end in view—viz., the effective training of his mental powers and moral principles.

The foregoing considerations fully confirm the general principle, already enunciated in a former chapter, that in infancy the knowledge of external things depends upon the material objects, and the ideas of actuating principles are imbibed from the personal example and conduct exhibited in the environment.

Considering, then, the early and effective operations of the infant mind, and the influential ideas that are then imbibed, it follows that the mental and moral atmosphere of the environment should be purged with not less care and solicitude than are observed in regard to physical and sanitary conditions. The culpable indifference and improvidence which have often proved so fatal to human life in regard to these conditions have elicited legislative interference for their enforcement in all large communities. The same culpable disregard of the mental and moral influences to which the human mind is generally subjected in its infant and helpless stages is productive of even worse results upon the progress of the human mind; but there is, unfortunately, no legal remedy for the abuses and demoralization so produced. Government interferes in the way of education at a later stage, after the preparatory processes have been already too effectively completed, and the germs of mental and moral corruption have been sown. The only available remedy is by means of the dissemination of correct principles, founded upon a knowledge of the working of the infant mind.

Natural affection, and the emotions thereby enkindled, afford strong motives for parental care and solicitude; but these are of themselves the offspring of the mental nature, and do not supply that pure moral atmosphere so necessary to the cultivation of the infant mind and to the infusion therein of true moral principles. Natural affections differ very little in people of all grades and varieties of character. Even the lower animals show great tenderness, patience, and indulgence to their young. It is remarkable that they rarely ever manifest anger towards them or use violence. They endure the most persistent provocation from their too playful or too exacting progeny with forbearance such as is not common amongst human parents. The ill-tempers exhibited by the lower animals, of which some instances are cited in the Appendix, are the result of ill-treatment received, not from their own parents, but from the hand of man.

The family institution is a form of environment eminently promotive of good influences on the young, by means of the exhibition of domestic virtues, stimulated by natural affection, on the part of the elder members. Experience of life, knowledge of men and things, and example of conduct are there practically manifested in the little epitome of the outer world to the younger members, who, in their earliest stages, imbibe the ideas so suggested, in the same unobserved but effective manner in which they also acquire language at the same period.

The imitative power and propensity are also there stimulated, and hence particular gestures, habits, and manners, mental and moral, are repeated in children, and are commonly attributed to inheritance, because no other origin was observable. So far, however, as these imitations tend to evil, the results are shown in chapter xv. to be the natural effects of more immediate causes, and the responsibility for their appearance is fixed accordingly.

The employment of ignorant persons in the nursery, and the delegation to them of duties properly belonging to parents, are common errors of domestic practice, and are responsible for some of the worst results of the prevailing custom of the middle and upper classes. To such example may often be attributed the tendency to falsehood so common amongst children, and many other ideas, superstitions, prejudices, and mischievous fictions that infest childhood, and often persist through life.

Another source of mischief, and even demoralization, in family practice is the habit of speaking and acting in the presence of very young children as if they were unobservant and had no understanding. Far from escaping their notice or being unintelligible, they are often attentively noted, for children are sometimes most intent when they seem most unconscious. Hence it happens very frequently that children are punished for knowing what was plainly said or done in their presence, and for imitating naturally and innocently what they have seen their tutors do. Disclosures of an unexpected and unwelcome kind are often thus made, with painful results to others beside the child. Seeing that infants are so observant, as has already been proved, and that example is often so ill-adapted for their imitation, and that, in fact, they see and hear so much that was not intended to reach and pollute their minds, it is easy to account for all the strange tempers and ill-manners which are developed in later childhood, without need to have recourse to remote ancestry or other imaginary causes of their origin.

As we have had frequent occasion to refer to the general practice of regarding very young children as unconscious and unobservant of what is going on around them, we may here digress to consider the reason of so general and influential a mistake. It seems to be attributable to the fact that facial expression is not a natural, but an acquired habit, derived from observation, and practised by imitation. Before it is acquired, therefore, observers assume, from the absence of the customary marks of apprehension, that the mind is unconscious. Until the usual response of facial expression is acquired and practised, the actual expression may be, and often is, one of vacancy. The habit of regarding the countenance of persons to whom speech is addressed, in order to engage attention and note the expression of feature, is quite universal amongst adults, and the result of his words is clearly manifest to the speaker, who is guided accordingly. This manner of accounting for a vacant expression in young children is strongly confirmed by the incident related in Note 6 of the Appendix, where the absence of facial response was sadly misinterpreted, and led to painful result.

Conscientious reflection on the influence of example, and especially of its potency in early infancy, when no palliating explanations are possible nor any atonement practicable, and considering also the responsibility thereby imposed on tutors, it might well be asked, who is sufficient for these things? If, indeed, evil in example and error in fact were required to be excluded, the answer would be simple. The task would be superhuman. But no such obligation is laid upon tutors; nor would it be appropriate in a world in which evil exists for a purpose, and is to be overcome, not escaped. The element of evil is necessary to the training of the human mind and to the testing of the moral nature. Its exhibition in the world is the practical proof of its real character, which is manifest in its results, and could be neither taught nor understood, nor conceived otherwise. Its most revolting developments have therefore a purpose which could in no other way be fulfilled. Fallibility, imperfection, error, and evil are parts of the insignia of the probationary state of this life —elements of the work for which the present life is ordained.

It is not necessary that tutors should be faultless in conduct. The essential qualification for the due discharge of the onerous duty they owe to the infant training is that evil shall wear its own colours, and never be made to stand for its opposite. The rule of right and wrong should be inflexible, and apply alike to tutor and pupil. Whatever is wrong should so

P

appear, and be neither excused, nor palliated, nor otherwise made to seem different from what it is. As it forms the special business of life to choose between good and evil, *both* are set before the free agent. It may be humiliating to personal pride to admit being in the wrong, but it is far worse to be convicted of it. The next best thing to avoiding evil is to admit it frankly, and make such amends as may be possible.

Errors in regard to ideas of external objects are comparatively unimportant, and may be corrected. Even whilst they prevail they do not affect character, nor stand in the way of vital truth. On the other hand, errors affecting the principles of conduct or Divine truth bar the attainment of spiritual excellence and oppose the supremacy of the higher nature.

Of all the influences directly and indirectly dependent upon environment, example is by far the most important to the cultivation of the infant mind. On it depends the only possible ideas an infant can acquire of moral principle. Even when he is able to understand precept, it derives its meaning and receives its interpretation from the exemplification of moral principles in his experience.

And of all that example comprises or implies, consistency is the characteristic which most certainly inspires and sustains that confidence which gives force and effect to personal influence.

SICKNESS, INFIRMITY, &c.—The various effects produced by the numerous causes that would come under this denomination differ so widely that it would not be practicable to discuss them separately, or even to classify them in such manner as to consider them severally. Almost all forms of physical ailment affect the mental powers, but the tendency in general is rather to divert than to diminish the capacities. Those which restrict the environment tend to encourage the reflective faculties in proportion as the range of perception is narrowed. They also tend to exalt the perceptive faculty in its exercise on the fewer objects which come under observation. If the authorship of all the eulogies that have been pronounced on the beauties of Nature were statistically classified, it is probable that the most affecting, if not the most numerous, would be those of valetudinarians. The sensitiveness peculiar to bodily weakness may contribute to the high appreciation such persons usually have for natural beauty.

Children suffer, on the whole, more than adults, and that form of suffering which is most general, and which excites the most sympathy, is teething. As it occurs at a time when they are so helpless and dependent, and is in itself scarcely susceptible of any form of alleviation other than that which sympathy affords, it would seem to be ordained for that object. On no other occasion does an infant seek so anxiously for the solace of the maternal bosom. In the midst of his suffering he will there sink into a peaceful sleep. He is at such times most sensible of his own dependence and of her sympathy and love. His firm trust and her warm sympathy are typical of the sinner's faith in the Saviour's love and in the efficacy of His office.

ENCOURAGEMENT.—It is probable that mental progress in the juvenile mind in regard to all kinds of knowledge, but more especially to the development of correct ideas and habitual practice of principles, mental and moral, is in no other way so strongly influenced as by the stimulus of encouragement and success. Children universally manifest the warmest appreciation of such attention and aid, on the part of tutors, as evince an interest in their efforts, whether to understand a new idea or to effect a desired purpose. Sympathetic aid or encouragement never fails to elicit a marked response, and is never thrown away.

RIDICULE.—We deprecate on principle recourse to this very common form of appeal to feeling, because it is an irritant. It may not be safely practised amongst adults on that account; and it is at best a cowardly proceeding to practise upon a child a remedy too pungent for an equal. The natural effect thereby produced, and the feeling so aroused, are too familiar to need any illustration; but we were struck on reading a recent narrative with the effect of ridicule upon a company of aborigines, the Veddahs of Ceylon. Mr. Stephens, who recently spent some months amongst them, was moved to mirth by the performance of a Veddah who was acting. Inadvertently, and forgetting for the moment the sensitiveness of these natives to ridicule, he laughed. On the instant an arrow whizzed past him, and he narrowly escaped a terrible penalty for his forgetfulness.

In whatever terms ridicule may be characterized, it certainly does not come under the category of what is *good*. It is evil in whatever guise it may assume, and is therefore *not* a weapon to be employed for overcoming or counteracting any other form of evil. Its effect on children is mischievous, and sets them an example they are but too ready to follow, and which, though it may make them many enemies, will never in a lifetime make them one single friend.

CHAPTER XI.

ON THE HABIT OF THOUGHT.

The force of habit is so generally acknowledged, and indeed is so sensibly felt in individual experience, that no argument could make it more convincing than the testimony it affords of itself. Naturally it is most influential in the domain of Thought, inasmuch as it there becomes the parent of most other habits. By general consent it is called second nature, and it may be second in point of order, but it generally attains supremacy in importance by the direction it gives to the operation of the natural powers.

The influence of habit is universal, determining the direction of the currents of Thought and prescribing the form of the activities of both body and mind. These activities take their course from the very first along the lines of least resistance, or of superior force. However faint they may be at first, these lines deepen by degrees, and eventually acquire directing force. Thenceforth the future course of Thought or mode of action follows the prescribed lines. Thus what was at first but a slight inclination, the fine dust of the balance, becomes a dominating power. In the end it operates almost involuntarily and mechanically, superseding the action of the Will, which gives its tacit assent to the familiar force. Hence character becomes resolved into a congeries of acquired habits of Thought, and life represents the sum total of their products. Insensibly, but certainly, the modes of action become moulded into habitual shape ; the very forms of expression, inflexions of voice, and smallest gestures become characteristic, and would even identify the individual when his features could not be seen.

Habit being thus universal in its sphere of influence, and so predominant, it deserves a first place in the training of the intellect both in order of time and importance. The fact that the formation of habit commences with the very first exercise of the bodily and mental powers, obedient to whatever forces may incline them in a given direction, suggests vigilant attention of tutors to the first indications of inclination in infant mental activities. Current practice does not recognize this duty, but assumes that, until habits have become fixed and predominant, they need no particular care.

There is nothing in the dominating influence of habit that affects the principle of free agency, for it is in its formation subject to the consent, if it be not by the inclination, of the Will. And even after it has become established, the supremacy of the Will may by a vigorous effort be restored, and the habit thereby checked, or even in most cases eradicated. Still, such effort is not often aroused in adults, and could not be expected of children, even if they had intelligence sufficient to understand the need for it. In the infant stage, the duty of initiating good and restraining evil habits devolves upon their tutors, who are generally but little impressed with the need that exists for their studious and vigilant attention to this particular. Hence, many habits formed in early infancy require to be overcome in after-life by means of efforts which otherwise might have been much better expended. In the meantime, character may have suffered, the lower nature may have been fortified, and much ineradicable evil done. Of such are the common habits of untruthfulness and ill-tempers, the results of which are often lifelong.

Habits are, in their inception, weak and easily amenable to correction, but those which are contracted in infancy, and with which, therefore, we are most concerned, originate with the ignorance, neglect, or example of tutors, from whom nearly all the first ideas are derived, directly or indirectly. They should therefore be prevented by salutary conditions of nursery environment and precautions of conduct, and should not need correction. Still, when that is necessary, the suasive influence of affectionate restraint will prevail, especially if exerted judiciously during the period of conscious dependence of the infant. In regard to those habits which are not directly inspired by example, and which originate in the infant mind, the sphere of infant Thought is so restricted, and the lines of possible inclination, the motive influences, are so few, that such habits are easily under the observation and control of watchful tutors, who, therefore, are also morally responsible for them.

Temper, which is the habit of feeling towards other persons induced by the treatment experienced, is the particular habit which is the most important of those contracted in infancy, and is so influenced by feelings, and dependent thereupon, that it practically rests far more with tutors than with the infant himself.

Imitation, the natural impulse of the infant faculties in response to the actions and manners of tutors, is a powerful agent in the initiation of some habits acquired in infancy. Hence the frequent and remarkable resemblances of gesture and other observable habits to those of tutors. From these resemblances, the origin of which is so obvious, it may be confidently inferred that other less noticeable acts are similarly imitated, and with equal effect, notwithstanding their less obvious manifestation.

Though originating in single acts, habits do not assume their proper character as such, or become apparent to observation, until they are indicated by repetition. Hence, their origin is imperceptible, and the progress of their gradual formation is often unnoticed until they have attained at least such partial development as to be recognizable as habits. In adults, who seldom enjoy the advantage of being corrected or warned in such matters, habits often grow up insensibly, without the knowledge of their victims, who, not "seeing themselves as others see them," persevere in them, to their personal disfigurement. Such habits of gesture are, however, unimportant as compared with those less noticeable ones which affect character and impair the beauty and symmetry of a life. Yet these commence as obscurely, grow as imperceptibly, and become as fully developed in the domain of Thought as those in gesture or manner of action. Habits of gesture are easily seen, and may be readily corrected, especially in childhood, but those which affect the currents of Thought are often quite imperceptible, concealed in the secret of the heart, unknown, it may be, even to their possessors. One of Goldsmith's intimate friends once offered to tell him what he was thinking. Goldsmith accepted the challenge, and then discovered, by his friend's reply, that his habit of thought was better known to others than to himself!

Although habits of Thought may not be noticeable to observers, and may be even unknown to their subjects, the existence of such mental habits is nevertheless certain, as is proved by the evidence afforded by those more obvious ones which betray themselves to the eye. For it is not reasonable to suppose that the imitative propensity on the one hand, or the natural tendency of Thought to gravitate, so to speak, in the direction of greatest attraction on the other hand, would be partial in their operation and confined to selected spheres. As natural forces operating generally, these causes assuredly produce their appropriate habits, whether visibly, in gesture and manner, or imperceptibly, in mental peculiarities. In short, it is certain that whenever any effort, whether corporeal, mental, or moral, is made, that form or direction of the effort which involves the least exertion, most facility, or strongest stimulus will gain an almost involuntary preference. Hence, lines of direction are traced lightly at first, and are gradually deepened into grooves that determine the course of mental action. Such is the natural history of the inception, growth, and development of that most potent element of mental and moral culture, habit.

The fact of the insensible origin and unobserved growth of habits obscures their real causes; wherefore, as these are not recognized, others are conjured up to account for them. Hence, when they become evident by their manifestation, they are generally attributed to natural depravity or to natural endowment, according as the particular habit may exhibit some infirmity of temper or some form of mental power.

Insensible as are the beginnings of the early history of habits of thinking, their influence is far-reaching and of the gravest import, as will be shown in the sequel. Hence, the conditions of environment, which determine the subject-matter for the earliest exercise of the infant mind, and which, therefore, influence the form and direction of mental effort, merit the most earnest solicitude of tutors, who in that helpless period of their pupils' history are responsible for those conditions. Equally necessary is it in after-life to beware of such slight preferences as might grow insensibly into the form and force of bad habits, and, on the other hand, to cultivate such as have a useful and beneficial tendency. These may become, under the fostering influence of encouragement, noble traits of character, and may prove powerful aids to virtue. Such, for example, are small acts of charity and self-denial, the fruits of which may develop into glorious deeds. Of such are many splendid endowments which abound throughout this country. On the other hand, the beginnings of self-indulgence, whether in the use of stimulants or other form of selfish taste, are maleficent, destructive of moral character, and make the thinking faculty the slave of the lower nature.

The order in which those habits are formed in infancy which are most influential in after-life is regulated naturally by the corporeal and mental conditions of that period. Each of them has, in fact, a natural and appropriate time when its inception and early growth may be easily insured. The recognition of these golden opportunities is an element of primary importance in mental culture, and one, therefore, to which the force of example and adaptation of environment should be specially directed. On the use of those opportunities it mainly depends whether successive stages afterwards be characterized by peace or conflict, and whether the higher or lower nature shall prevail. Their neglect or abuse in the first two periods determines whether the period of childhood be one of perpetual trouble and collision or be marked by peaceful progress—whether the youth proceeds to school with a disposition to learn, and with those qualifications most conducive to a prosperous career, or whether such adverse conditions as idleness, distaste for learning, and an intractable temper oppose the efforts of the master and the progress of the pupil.

The multiform developments these and other allied characteristics assume in adults need not here be discussed. They are present to the view of every individual, as seen in others and felt by himself. Reflecting on them and their early origin calls to mind the speech of a celebrated French criminal, who, from the scaffold, anathematized his parents for their neglect of parental duty in his childhood, whereby they had become the authors of his ruin.

In so far as action is influenced by temperament, habit will of course be modified by this important factor, which therefore is in all cases to be taken into account. Although it does not affect the moral influence of treatment or example, temperament modifies their force and effect very materially. A natural tendency to activity in one and to comparative indolence in another, the sensitiveness of one and the callousness of another, and other such-like differences arising from temperament, will be influential accordingly. Such differences will suggest to tutors the manner of their proceeding, and could not be here discussed with advantage. The rule of conduct applicable alike to all cases is prescribed by the fact that the natural forces in operation produce their appropriate natural effects, whether in mental and moral or in physical nature.

There is one principal habit which materially affects the general operation of the mental faculties, and which therefore has a decisive influence on all other habits of Thought. Indeed, it is a chief exponent of character. Whether the mental faculties are habitually under the control and direction of the Will, or are generally allowed to follow in a desultory way whatsoever course outward objects or internal suggestions may prove most captivating, indicates that particular habit or condition of the mind which determines the general force and effect of its powers. Reverting, for the sake of illustration, to a simile employed in a former chapter—the castle may either be kept vigilantly and under control, or be unguarded and open. Its inmates may either be under discipline or left free to take their ease and follow their own devices. Admission may be either by licence, or indiscriminate. The gate may either be closed to all but those visitors who have a useful mission, or it may be left ajar for any loiterer, casual, or enemy to enter at his own list.

Such being the distinction between the two leading principles on which the mental powers are exercised, one or other will, in every case, be predominant in a greater or less degree, and will constitute the mental characteristic of the individual. One or the other will determine the general or habitual mode of mental action. Neither will, even in the most extreme case, prevail *entirely*, for no one is either so utterly and helplessly subject to casual influence as to be unable by strong effort to exert some measure of control; nor, on the other hand, could any one obtain and exert at all times such supreme command of his thoughts as to resist all external forces and the occasional pressure of overpowering circumstance. In either case the Will may be exerted: in the one, naturally, and with the prevailing force of habit, and in the other by a painful, unaccustomed strain, of doubtful issue.

Of these two states or habits of mind, one gives force and efficacy to the mental powers and strength to character; the other weakens the intellect, and tends to other habits which subordinate the higher to the lower nature. The dissipating effect of a listless habit extends to every department of mental action. The habit of controlling and directing Thought to whatsoever may be most worthy of attention, or may most need it, engages the whole force of the intellect in what is most useful or most necessary, to the exclusion of all other claims or attractions; whereas the opposite habit reverses this order, and tends to divert attention from duty, culture, and salutary Thought to the more seductive subject-matter which is ever suggested by passion, appetite, and selfish inclinations. Hence the former habit is to be sedulously cultivated and the latter as jealously avoided.

These two habits have a certain relation to those

popularly described as thoughtful and thoughtless; but these terms do not fully convey our meaning. The habit of listlessness or laxity which we deprecate is not necessarily one of thoughtlessness. It may be characterized by much desultory, unprofitable, or demoralizing Thought. It does not imply a want of Thought, but of its useful and effective application. Uncontrolled and wandering Thought is generally led into unwholesome currents, and exposes its victim to the worst forms of temptation, whilst at the same time it effectually weakens the power of resistance. It may be, and too often is, engrossed by captivating ideas, of which the lower nature has always an inexhaustible supply at hand. The term *listlessness* is the mildest name by which we can distinguish the habit in question.

By inviting and encouraging the assaults of temptation and weakening the power of resistance, listlessness is especially adverse to the development of moral principles, and straitens yet more narrowly the too narrow path of virtue. But, though its worst and most powerful influences take that direction under the dictation of the lower nature, its operation is universally adverse to mental action. It affects the memory, judgment, and all the reflective and reasoning powers. Its victims complain of their inability to collect their thoughts or to concentrate them on any given object. Habituated to wandering, their thoughts are not readily restrained, but pursue their natural bent in spite often of the strongest efforts of which the crippled powers of resistance are capable.

On the other hand, the habit of controlling the action of the thinking faculty, and directing the currents of Thought to present purpose, confers great mental power. The individual self is truly its own master in all those respects wherein the opposite habit enslaves and makes him a creature of circumstance, a prey to his lower nature. The ruling motive of the strong-minded man who maintains a mastery over the direction of his mental faculties is determined by the principles he has imbibed, and not by the feeling of the moment. He is a reasonable being, not only as possessing the full measure of reasoning powers, but because they furnish the actuating motives which determine his judgments and regulate his conduct, freed from the enslaving influence of adventitious circumstances.

When Thought is left to the influence of natural gravitation, and its possessor virtually abandons the government of his mental estate, those ideas and external influences, and those internal suggestions of appetite and passion, which are most powerful and obtrusive, will captivate and engross it. Listlessness is therefore a condition the most unfavourable to mental attainments. Concentration, accuracy of observation, clear ideas, memory, and the reasoning powers, the necessary qualifications for mental work, are all impaired by the superficial and desultory habits to which listlessness tends.

Attention and concentration are difficult to command, even by those who have most complete mastery over their faculties, and are therefore, above all, those elements for which listlessness most disqualifies the mind. The superficial and casual habit of observation to which this state of mind naturally leads does not afford the materials necessary for the formation of clear and definite ideas. Those they supply are naturally imperfect and unreliable. Hence the data on which memory and the reasoning powers are exercised are unfit for the purposes of high mental attainment. The critical accuracy and precision which are so important to effective mental work are incompatible with a desultory habit of Thought, because they depend on the exercise of exactly opposite qualities. It follows that persons who have acquired this loose habit are deficient of mental attainments in proportion to the degree to which it prevails.

A sense of deficiency is generally felt by those persons in whom the habit of laxity and listlessness has been most prevalent, but is rarely, if ever, traced to its proper cause. As the habit most probably originated in early childhood, the victims may be themselves ignorant of the time and manner of its beginning, and it is therefore usually, almost universally, attributed to natural infirmity or inherited defect. Hence, they regard as special gifts the powers and facilities possessed by their more accomplished neighbours. They fail to perceive that the mastery these have acquired, and the control they possess, are the result of a careful and diligent use of their faculties exercised under conditions which were neither inherited from parents nor conferred by natural endowment.

The habit of regarding acquired faults as special and personal inflictions, and acquired attainments as special gifts and favours, is indulged by many persons, as the theory supplies a convenient excuse for many grave imperfections of character and conduct, and for transferring from themselves the responsibility they would otherwise imply. It covers a multitude of sins, and is therefore a favourite doctrine. It is not, however, to be confounded with the fact of diversity of natural power. Minds undoubtedly differ in power and activity, as sheets of paper differ in size, texture, and sensitiveness. The doctrine we deprecate implies that, besides the differences of capacity, there are also ready-made inscriptions on the blank sheets! This subject is discussed in a succeeding chapter on Inheritance, but is here alluded to as being one of the numerous progeny of the habit of mind now in question.

The theory of inherited Thought paralyses effort by making it appear ineffectual, and encourages a *laissez-faire* habit by which it is so often suggested. It gives rise to feelings of jealousy of attainments which would otherwise inspire emulation. It is opposed to philosophy, morality, and Scripture. The same Authority who declares the different number of talents committed to men's stewardship emphatically denounces the neglect of the *one*, and requires that it should be employed in the same manner as the five or the two. If the theory affected the acquisition of mental attainments only, it would still be worthy of much consideration, but its worst results are manifest in the conduct of life. There it is urged to excuse the neglect of unpleasant duties, especially those involving the control of the lower nature, and it thus conflicts with the best interests of mankind and the ordinances of Providence. It raises a defence against the dictates of conscience and the Divine command. It fosters the idea that the evils born of lust within are imposed from without by foreign influences. It makes men consider themselves the victims of evils of which they are themselves the real authors. We have met with persons who pleaded their inability to resist the temptation to drunkenness and even to lying on the ground that the weakness was inherited! The proper remedy for all the evils consequent on a lax and listless state of mind and habit of thought is the counter-agency of self-control and effort. In so far as man is not his own master by the effective control of his thinking faculty, he is a slave of many masters, comprising in their number the greatest tyrants of the lower nature.

The foregoing remarks on the nature and general tendency of the habit of controlling and directing mental action, and of thus preventing a desultory and passive habit of allowing Thought to wander in whatsoever direction it may be most attracted, will establish our proposition as to the importance of cultivating such control, and of sedulously guarding against the opposite habit in the early training of the mental powers.

The fundamental principle which determines whether the mental powers shall be directed to definite purpose or be left free to pursue a desultory and disorderly course, is discussed in the next chapter, on "Control of the Thinking Faculty." This is an exercise of the Will in the first instance, and continues so to be until the functions of the Will become superseded by the force of habit, or are crystallized into that form. The general management of the infant Will should be effected when, in its very earliest exercise, it has no force of its own, and when, therefore, it may be moulded into shape. This subject is elsewhere discussed, and we are now concerned particularly with its bearing on the formation of the first and fundamental habits of Thought.

The particular exercise of the Will in mental action takes the form of *attention*, which is, in fact, a directing of the faculty in response to some kind of sensation. Thought is thus first diverted from the absorbing influence of personal feelings by external objects which excite interest and invite attention. This responsive action is evidently a natural exercise, as is proved by the voluntary attention which must be given for the acquisition of language, the common possession of all infants with normal capacities. It does not appear to be necessary, therefore, to sow the seed of the habit of attention, but only to encourage its cultivation, and give to the natural tendency the form and force of a confirmed habit. Aid and encouragement, the chief inducements to infant efforts of all kinds, will tend to confirm and strengthen the natural impulse, whilst neglect, discouragement, and repression will have the opposite effect. Thus, when the infant mind is intent in observation, the effort so manifested will either excite the sympathy of tutors, and be aided and encouraged, or it will be neglected or interrupted, as though the workings of the infant mind were of no moment and unworthy of notice. In such ways is the growth of the natural tendency stimulated on the one hand, and confirmed into the form of habit; or, on the other hand, new seed is sown of a nature to impair, or probably supersede, the salutary and natural exercise, and to substitute one which will subject mental effort to the influence of desultory forces.

The influence of the material as well as of the moral environment in the first determination of the prevalence or subordination of mental control and of the mode of treating infant effort must be evident from the nature of things. The vigilance and attention on the part of tutors which would be exacted for the due discharge of duty in this respect are no doubt a serious consideration, but they are, in practice, much less onerous than they at first appear, and they save an immeasurably greater trouble and difficulty in the later periods of childhood, and more particularly in that stage of school-work for which all the previous ones should be regarded as preparatory.

The initiation of a habit of attention in the period of babyhood may be fostered by encouragement given whenever baby is noticing any attractive object, for which, whilst he is in arms, tutors have excellent opportunity. In their dependent state, babes unavoidably absorb a great amount of tutors' personal attention. Happily and providentially so, for the purposes appropriate to that stage of their physical and mental progress require such attention, and there is no occupation so fascinating to a thoughtful tutor, nor any other so useful and beneficial to the babe, as that of watching, entering into, sympathizing with, and encouraging the first intelligent efforts, the first buds of future thought. By universal consent, the baby-form and expression of feature are accepted as the material emblems of the spiritual, angelic ideal, and there is certainly no object in the whole round of Nature that can be compared with those characteristics for innocent loveliness. No other conditions could so effectually invite and win the attentions of tutors, nor could these attentions be in any other way so beneficially exerted as in the work of developing those buds of mental habit, which, when fully expanded, will either impart to the future life a higher form of beauty or disfigure it with mental deformity.

In the later stage of infant life a habit of attention in mental exercise may be even more effectually aided and encouraged by vigilant improvement of the opportunities afforded by the active tendencies of mind and body characteristic of that period. The methods adopted in the Kindergarten system are remarkably effective in engaging attention and exciting interest in the infant mind. There the "gifts" are made as attractive as such lifeless, mechanical, and unsuggestive things can be made, and what is deficient in the "gifts" themselves is compensated by the suggestions and attentions of the teachers. These are, in fact, the effective part of the system, and would serve their purpose with almost any form of material appliance. In so far as the system succeeds in engaging attention, and in making the exercise of the natural mental powers habitual and easy, it has our strong approval, though we do not equally sympathize with the direction given to the currents of Thought so created. The fanciful and unreal nature of the ideas suggested by the teachers from remote resemblances arising out of the permutations and combinations of the "gifts," as well as the mechanical character of the ideas themselves, seem to us to beget a habit of substituting imaginary characteristics for the substantial realities of things.

The inquisitive habit of childhood suggested by the stimulus of speech is peculiarly favourable for encouraging and confirming a habit of attention. In this stage of the mental progress tutors have no difficulty in perceiving the drift of their pupils' desires, in which their opportunity consists, or in availing themselves of the occasions so arising. They may then most effectively enter into, and sympathize with, their pupils' proceedings, and give the lead to effort or confirm and aid its success. In proportion as they thus give a lead or encouragement will the interest and attention of the pupil be attracted and won. And, along with it, the trust and confidence of the pupil will be equally developed.

By such natural and rational means the mind may be gradually provided, during the early stages of infancy and childhood, with habits suited for the work of the succeeding stage of school-life. By the time a youth enters upon his educational career, his mental habit has generally become formed, and has assumed a shape which is not easily altered. The most influential of all mental habits—that of controlling, or neglecting to control, his thinking faculty —will certainly have been initiated in one way or another, and all other mental habits will take their character from that one in a greater or less degree. The work of education tests the habits that have been previously formed, and its issues depend far more upon them than upon any subsequent ones that may be created by school discipline, however wisely and judiciously these may have been inculcated. It would often be an unspeakable blessing to both master and pupil if the formation of habit had then to commence; but that is not possible. The mind, if not trained in a right direction, will obey other impulses and receive from them such shape as they may impart.

The facility with which some scholars seem to

learn without any very apparent effort is not always due to superior power, but sometimes to habit of controlling and concentrating their thinking faculties. He who possesses this inestimable power, which can only be acquired by cultivated habit, has a great advantage over a fellow-scholar who, though perhaps of greater ability, has to contend against the multiform innovations of casual thoughts. These must either be ejected or their confusing influence must be endured as a fatal hindrance to the study properly in hand. To one, the appearance of a butterfly or a bird would be so captivating as to completely distract his attention and cause an interruption, with dissipating effect. Such a one has no chance in a competition with a neighbour whose attention is fixed on the work in hand to the exclusion of all such interlopers. The place in class in such cases is not determined by the relative abilities, but by the force of habit, of the pupils.

The too general practice which defers the sowing until harvest-time is the reason why a crop of weeds so often encumbers the ground which the schoolmaster is employed to clear and to cultivate. The mental habits contracted and the body of ideas collected by youths in general during their early childhood, before any training was considered necessary, show how large a proportion of what was then done has to be corrected, counteracted, or extirpated; or, failing the success of these reforms, how their lives will be disfigured and their mental progress impaired.

Habits of *observation* are generally formed very early in life, not only as regards the habit of noticing things, but the manner in which observations are made. Whether a child acquire a habit of noticing things generally, or of neglecting to notice any but those which are forced upon his attention, is usually determined at a very early stage, and depends mainly on the extent to which tutors encourage observation by attractive devices and interesting explanations. There are many who altogether neglect all attempt to excite the interest of children in any of the things about them, and many others who discourage the natural disposition children evince in the objects of their environment. The number who improve the numerous opportunities of encouraging an observant habit is comparatively small in proportion. Hence it is not common to meet with a really and intelligently observant child. It is to be hoped that a time may come when the early mental training of children may be regarded as an art for which some qualification shall be deemed necessary. The practice of relegating the care of children to persons chosen without any regard to their ignorance of everything they ought to know is productive of much mental degradation, and is an effective hindrance to the progress of mental culture. The fact that the infant mind is subject to the operation of natural laws, and that it is moulded in accordance therewith, is both as true and as deserving of the attention of scientists as are the economic laws which regulate the price of bread and cotton.

The *manner* in which things are observed becomes also more or less habitual in childhood, whether casually or carefully, generally or in detail, and also whether the evidence of the ear or that of the eye prevails. The habitual mode of observation influences materially the character and value of the ideas formed and the use that may be made of them. Inferences and judgments depend necessarily upon the accuracy and sufficiency of the ideas or data from which they are derived. Moreover, things are necessarily remembered, not according to what they really are, but as they were regarded. It is pitiable to witness the embarrassment of some persons in their futile endeavours to describe what they have seen or heard, owing to the partial or cursory way in which they had noticed the sensations and formulated their ideas of them. Attentive observation is essential to a definite idea, and, if the idea be indefinite, memory cannot supply the original defect.

In this connection we may remark upon what seems to be a natural impulse of the imagination in regard to misty, indistinct, or imperfect images of objects seen or sensations of sounds. There is a tendency to supplement and complete by the imagination what was wanting in the evidence of the sense. When, for instance, an indistinct object, like a nebula, is perceived, its want of definite form is generally supplied by the imagination, and a general outline is manufactured. Hence, when a powerful instrument is employed, the real figure as thus disclosed may be very different from the one imagined or fancied. A person who during some months suffered from very dim vision made several acquaintances during that period, and saw their features very indistinctly. Still, the images in his mind were definite. On the restoration of good vision, however, the images so distinct in his mind suffered astonishing change—such, in fact, as seemed almost impossible. In like manner uncertain sounds receive imaginary interpretations, "and as the bell clinks, so the maid thinks." The unreliable nature of supplemental imaginings has probably a good deal to do with some of the strange and unaccountable misconceptions into which adults of strong understanding frequently fall. And they have an important bearing on the habit of formulating ideas, either casually or upon insufficient data, for such ideas, being imperfect, if not positively incorrect, must be more or less misleading.

As all reasoning is based upon reflection, and therefore involves the evidence of ideas previously formed, it follows that all reasoning processes must be vitiated, to a greater or less extent, by any imperfection or error in the ideas on which they are exercised. Habits of observation, therefore, are of the utmost importance to sound reasoning and correct opinions.

Though the influence of habit is not very apparent generally in matters of opinion and judgment, it is nevertheless often powerfully exercised therein without being perceived. The mode of observation, the estimation of evidence, the very attitude of the mind towards the subject-matter submitted to its judgment, are all subject to the influence of habit, and all contribute to the result. A casual observer sees things differently from one of more discriminating habit, and his reasonings and inferences will necessarily differ accordingly. In all matters which require close and accurate discernment, a thoughtful man's opinion will differ materially from that of a man of thoughtless habit. One who gives his attention habitually to the observation of details thinks on different lines from one whose habit is to take comprehensive views. The former will note the colour of the hair and eyes, the particular form of nose, and other such details of feature. The latter will consider and recollect the general expression, behaviour, and other characteristics of the "tout ensemble," with little regard to those particulars to which the former gave his almost exclusive attention.

The mutual relations of master and pupil depend to a great degree upon the habits with which the latter enters upon his school-life. His temper, disposition to learn, power of controlling his thoughts in the way of attention, and habit and mode of observation are all influential factors, which will have become, in one way or another, habitual, and will determine the attitude of the pupil towards his teachers and the issues of his school-work. It is therefore essential to a right adjustment of effort on the part of each that these habits should be known and taken into account. In the case cited in chapter xv., in which a pupil, after two years' trial, was deemed by his teacher to have no ability for learning music, the event proved that there was no deficiency of power on the part of the pupil; on the contrary, unusual power was afterwards displayed, nor was there any indocility. The teacher had failed to ascertain the child's mental habit, and therefore failed to bring his powers into effective operation. Similar results sometimes occur where habitual reliance on one of the senses has given rise to habitual inaptitude or slowness of apprehension by the other on the part of the pupil. In such cases the inaptitude suggests to an observer the want of natural capacity where habit only may be the cause.

Seeing that almost every act of life, whether physical, mental, or moral, is the exponent of some habit by which it is moulded, *character* is a summary of the collective habits. Thus a man is characterized as thoughtful, kind, patient, virtuous, or careless, harsh, petulant, immoral, according to his habitual conduct. In each case reference is made to general habit, and not to particular acts which may have been dictated by special occasion and be of exceptional nature. Considering also how insensibly Thought may be diverted and imperceptibly acquire the force of habit, jealousy of the indulgence of inclination is appropriate to every period of life. Though the foundation is laid in childhood, and the principal lines are then determined, the edifice of collective habit, which constitutes the character, is always being modified through the whole term of life.

Absence is a state of mind in which, for want of control, Thought is so absorbed in reflection that external sensations appeal in vain to attract the perceptive faculties. To one who has fallen into this unhappy habit, purity of mind and Thought may make it comparatively harmless to moral nature; but that is rarely to be expected where such want of control exists. In any case, it is inimical to mental progress, and inconvenient both to its subject and to those with whom he is associated. As a distracting influence it is even worse than that of external objects, as its inciting causes are always present. It is a snare that besets every situation, and is a serious hindrance to mental progress, especially in youth.

In its bearing on moral principle and the relation of the lower to the higher domain of Thought, habit assumes its most important aspect and exerts its greatest force. The suggestions of the lower nature are ever present and always powerful. They constitute the great proportion of those unbidden and casual intruders that are so ready to enter into and possess the unoccupied mind. And when they obtain the supremacy in the castle, their force increases with indulgence inversely as that of the resisting power. Hence it follows that lustful ideas, brooding over real or imaginary injuries, revengeful thoughts, vain lamentations over irretrievable losses, and other such currents of reflection by which evils are multiplied and magnified are sure to follow in the wake of the invading enemy. Incapable of conferring any good or answering any useful purpose, they are nevertheless most influential in embittering the cup of life and imparting strength to temptation. They are the active forces in the warfare of the lower against the aspirations of the higher nature.

The nature of a man's occupation exerts a material influence on the currents of his thoughts; there is therefore a certain intimate relation between the daily employment and the habit of thinking. Hence, certain habits tend to qualify and others to unfit a man for certain avocations. We knew a doctor, an excellent man, who had been most unhappy in his choice of a profession. His rash, impulsive habit of thinking was eminently unsuitable for such a pursuit. A stranger, suffering from a violent attack of colic, on his arrival at the town where the doctor practised, not knowing of an apothecary to supply the simple remedies he required, sent for the doctor, who, in spite of all the patient's protestations, insisted that the case was one of acute inflammation, and prescribed accordingly. The only part of his prescription which was of any use was the apothecary's address. The next day he would not believe the waiter of the hotel, who told him his patient was at table eating a hearty lunch, and could hardly be persuaded of the fact, as he insisted that his patient was seriously ill. Such cases of incompatibility of habit with occupation, though certainly not common, are sufficiently so to suggest the consideration of habit in the selection of a career for a youth whose habits have become pronounced.

It stands to reason that all those pursuits which bring men face to face with Nature in her beauty and purity must have a better influence on habits of Thought than those which expose to view the degrading indulgence of passion and appetite—those which subdue the energies of nature to the praise of their Maker, rather than those which reduce the master-spirit of the higher nature into unholy subjection to the law of the lower. These spoil Nature of her bounties to devote them to unlawful abuse, whilst those obey the Divine injunction, and place man in his own appropriate relation to the beneficent designs of the Almighty.

The engineer, who applies the forces of Nature to relieve mankind of drudgery, and to free their faculties for higher uses; the manufacturer, who utilizes and turns to economic account the products of the earth; the merchant, who interchanges the commodities of different countries and distributes the bounties of Providence; the poet and painter, who, in studying Nature, create art; the schoolmaster, who stores the young mind with useful knowledge and wholesome principles; the divine, who points out

God in all His works of nature and of grace, who holds up to the gaze of the pilgrim the Cross, with its Crown high up above all that is earthly or sensual —these have all salutary subject-matter supplied for the exercise of their thinking faculties such as could scarcely fail to beget habits of Thought in keeping with the nature of their avocations, excepting to those who cling to the lower and baser elements instead of ascending to the sublime and spiritual aspects in which God is revealed.

The force of habit, and the sway it exercises over the mind, are sometimes such that habits in themselves good may become snares by unduly absorbing time and attention. Devotion to a particular pursuit, and making it a speciality, may thus beget an exclusive habit, more particularly where enthusiasm prevails. In such cases duties may be neglected because of the engrossing nature of the special pursuit. A man may thus, in becoming a profound philosopher or in accumulating wealth, neglect his duties as a husband or a father to a greater or less extent. His habit in this case, though lawful in itself when duly regulated, may become a snare. In other cases, enthusiasm in a particular pursuit may so exaggerate its relative importance in the estimation of its subject as to make him disparage the value and importance of other pursuits.

CHAPTER XII.

ON THE CONTROL OF THE THINKING FACULTY.

THE power of controlling and concentrating the thinking faculty is the most important attainable accomplishment of the mind. On it depend the use that may be made of knowledge and the reasoning powers, the extent to which they are available, the measure of their efficiency, and the facility of their application to any given purpose.

The control of the thinking faculty is entirely subject to the Will, except when sensations are forced upon the attention by irresistible circumstance, or when the active exercise of the Will is superseded by habit. In other words, mental action requires the consent of the Will, either by direct effort or by passive assent, excepting only when the nature of a sensation or of its presentment is such as to force it on the attention. In the latter case, though the impression on the mind may not be resisted, yet the Will alone can determine any action to be taken thereupon. Its practical freedom and supremacy are not thereby impaired. Such impressions are of an exceptional nature, and need not be here considered. The functions of the Will in regard to the control of the thinking faculty will therefore be discussed with reference to their direct action, and to the passive assent involved in the crystallized form of habit.

This latter mode of operation prevails to a greater or less extent in every individual, the Will not being exerted by direct effort in regard to Thought or act separately, but being merged into a general and habitual groove. Hence the habit of controlling the thinking powers is the most important of all habits, inasmuch as it supersedes to some extent and subordinates the supreme power of the mental estate.

Seeing that, as has already been shown in a former chapter, habits may be so insidiously and unsuspectedly begun, and that they may insensibly acquire prevailing force, the most jealous watch should be maintained against the forming of any habit, or the submitting to any such influence, as would tend to weaken the supremacy or suspend the action of the Will in the domain of Thought. A loose and careless habit, so easily acquired by mere neglect, and so often lapsing into that state of listlessness described in the last chapter, tends to establish a habit of neglect, and thus to weaken the power and suspend to a greater or less degree the operation of the Will generally.

As the mastery of the mental powers is the greatest achievement attainable, it is also by far the most difficult to acquire and maintain. The difficulty of exercising full command of the thinking faculty is often felt severely by, and causes a great strain to, even those who have practised long and striven hard to establish and keep habitual mastery of their mental powers. If, notwithstanding the vigorous exertion they apply to the task, such persons find the force of passion, strength of appetite, or pressure of circumstance nevertheless so prevalent, how must they be captivated and overpowered who, unaccustomed to such effort, and habitually passive, are comparatively unresisting?

Hence it follows that in mental training a first place in order of importance must be conceded to the attainment of a mastery as complete as can be acquired over the exercise of the faculties in order that effective control may become firmly established as a settled habit and the keys of the mental castle may be in possession of the rightful owner. So fortified, the full force of the mind may be concentrated upon, and the entire stock of knowledge summoned to, any given subject in preference to, and to the exclusion of, every other. The gates of the castle locked against intruders, and the forces within its walls under command, the full strength of the fortress may be directed upon any undertaking. Neither embarrassments within nor enemies without may then interfere with the mental operation. On the other hand, if the mental forces be divided between disorder within and enemies without, weakness, vacillation, and defeat are as certain in this event as decision, firmness, and triumph are in the other.

The Will, being the power by means of which the operations of the mind may be controlled, demands special notice in this chapter.

Every sane person is conscious of the freedom of his Will—of the power, that is, of acting or of declining to act in any matter within the limits of his physical or mental powers. As regards the mysterious nature of this wonderful agency, and the mode of its operation, nothing certain is known; but of the fact itself all is known that is necessary to a right exercise of the vast trust it imposes, and to turn it to such account as may best serve the purposes of life and the cultivation of the mental and moral nature.

The Will, being a natural force exerted upon the material organism, is necessarily distinct therefrom. It stands to reason that a force cannot act upon itself. Whatsoever forces actuate a material body must be distinct from that body, even though they may be inseparably joined thereto. The electric current which makes a magnet of a piece of iron, for example, is not inherent in the iron, but is nevertheless an inseparable component of the magnet, which is such only so long as that connection is maintained. The fact that certain animated organisms of a low order fulfil the functions of life entirely by involuntary action, without any power of volition, and also that some of the vital functions in man are involuntary, affords strong presumptive evidence that the Will is separate from, and not inherent in, the organism. The dual nature of man, consisting of a material organism and of forces natural thereto and acting thereupon, seems to be thus established. This is clearly taught in Scripture. St. Paul describes a law of his corporeal members "warring against the law of his mind." Whatever may be the essential nature of the connection, however, the relation of the body and the mind to each other is clear. They cannot be confounded together any more than the material elements in Nature may be confounded with the forces that operate upon them, or with the laws which regulate those forces. Practically, there is, as St. Paul expresses it, a law of the members and another law of the mind, and these laws or motions of desire conflict whenever the suggestions of bodily appetite and desire oppose the moral law, the basis of which is the knowledge or discernment of good and evil. In this conflict between the moral law and the desires of the lower nature the highest functions of the Will are exerted, and a great part of its natural history is revealed.

The Will comes into operation in the very earliest stages of life, and is then stimulated by appetites, feelings, and bodily impulses, and it continues to be so actuated both in the human and in the brute offspring, except when it is attracted by the sensations of external objects and modified by the suasive force and vicarious action of parents. In the lower animals the Will is actuated through life by the suggestions and impulses of the corporeal feelings and appetites only, but the human infant is subject to a higher law, to which his lower nature, with its appetites and feelings, is to be subservient. The first ideas of the moral law are suggested by the example and conduct of tutors, and are imbibed by means of the imitative propensity and by inference. These ideas thus slowly acquire motive-force, and are afterwards developed by exercise in their proper domain of regulating conduct through all the later periods of independent action. The power the human subject thus acquires in subjecting the lower nature to moral principle determines the measure of his departure from, and superiority to, the type of the brute nature.

There is in all living creatures a natural impulse to exert the powers they possess. In the lower animals, which have no moral principles to imbibe, the bodily faculties are developed quickly. Their scions speedily attain a condition of comparative independence. In the human infant it is otherwise.

In him the period of helplessness and entire dependence is long protracted, and thus affords to tutors the opportunity and conditions necessary for the training of the higher nature, and especially of the Will, its executive power.

The movements of the limbs seem to be at first involuntary, unmeaning, and uncertain, but they soon become more decided and expressive as the Will becomes stronger by exercise and the limbs themselves gain power by use. Appetite and feeling, of whatever kind, stimulate the Will of the young creature to the exertion of such powers as he may possess; but so long as the period of helpless dependence lasts, the Will, the key of the higher impulses, is subject to the moulding influence, suasive force, and vicarious action of tutors.

The birth of the Will occurs in these earliest motions, which are involuntary in the beginning, and become gradually amenable to the growing force of the Will, which is developed by exercise. Its growth proceeds slowly, but certainly, in whatever direction the suggestions of feelings and appetites, modified by the influence of tutors, may determine. The direction as thus initiated, whether by the unrestrained operation and stimulus of the lower nature or as these impulses may be modified and directed by the superior influence of moral suasion, is that in which development proceeds and power of Will is acquired.

With the beginning of the life of each individual begins the education and development of all his powers. The limbs and organs of sense then commence their functions and uses, and the mental powers begin their eventful progress, taking from the very first such direction as the environment, including personal influences and the attractions of external circumstance, may determine. As each meal of food nourishes the body and does its part towards the growth of the physical system, so, in like manner, each mental feeling and experience, from within and from without, contributes its share and exerts its appropriate influence in the structure of the individual self, the Will, and all the mental and moral powers. Hence, the treatment and conditions which educate the brute would produce the same brutish nature in the human offspring, so far, at least, as the natural faculties admit of such development. That is to say, that such as are the forces in operation, such also will be the result. The Will, temper, and disposition are moulded by the influences under which the individual lives. It is not possible to escape the conclusion that nothing is lost, either in fact or force, that ever reaches the mind. It is not reasonable to suppose that the infant mind or sense could or would exercise selective choice as to the admission of one or the rejection of another sensation or kind of sensations. Each and every experience of pain, perplexity, or grief, and of example and treatment, does its part in educating the Will and in producing or modifying temper and disposition. The sum or balance of the whole is manifest in the resultant state of the mind.

It follows, from the foregoing remarks, that the Will of a babe is, for a time, in the power of tutors, and may then be actuated vicariously, not by any form of force, but by moral influence. The tutors' ministrations to the wants and indulgences of the desires of the babe, equally with the denials and restraints that may be necessary, should be effective but tender, firm but affectionate, and reasonable as if he were capable of understanding; for these are the lessons by which his moral understanding is first enlightened, and they are for a long time afterwards the only means by which it is educated and developed. These are the influences which originate those faint traces which deepen into lines, and at length form the grooves in which the Will is habitually directed.

The directing and correcting function of tutors in regard to the infant Will is a first and most important duty, and should commence with the very first indications of Will manifested. It is an important step in the "nurture and admonition" on which the ultimate character so much depends. By the term correcting, we do not imply any form or degree of punishment, than which nothing could be more inappropriate to the age and condition of the babe. Unfortunately, the term, when used in relation to children, is generally understood to imply punishment. The subject of governing by a system of rewards and punishments is discussed in chapter xvi., but need not be further noticed in this place. The duty of directing and correcting the early indications of the infant Will cannot be begun too soon nor be exercised with too much tact and judgment, as it is the foundation of all subsequent training. The feeling that treatment excites, whether on a given occasion, towards the actor, or habitual, the result of experience towards persons in general, regulates the attitude of the subject-person. And whatever the feeling, whether temporary or habitual, the Will is the executive power that gives it effect, so far at least as it is not influenced by other considerations or forces. Hence the Will and the temper, both resulting from treatment, are intimately related both in their origin and operation, and are the two factors first developed in the infant mind. Of the two, the Will, as being the executive power, is the more important as regards the control of the mental faculties and the conduct of life.

The education of the Will depends upon the discharge or the neglect of the duty just described. The effect of neglect is to leave the direction of the Will to chance influences and natural appetites. The result is represented by the strange variety and unaccountable developments that exist even amongst members of the same family. The results accord with the casual forces by which they were wrought. They are *natural*, not as being in any respect inherent in the individual constitutionally, but as being the natural consequences of the forces that produced them. The harvest is in kind like the seed which was sown in the spring-time of life.

The influences that affect the exercise of the Will in general are—(1) Parental suasion and example; (2) Physical force; (3) Hope or fear of consequences; (4) Human law; (5) Social amenities and obligations; (6) Moral principles; and (7) Divine command.

Of these, the first three are common to both the brute and the human races, and the last two are the exponents of that conflict between the sensual and the mental impulses which St. Paul describes, and with which every Christian is but too familiar.

1. Parental influence, as it is exerted by the lower animals, is never angry, seldom violent, and generally suasive and considerate. As exercised by the human

parent, it is generally more tender, but less consistent, and is, or ought to be, always tempered in such manner as to exemplify and suggest moral ideas and principles, for which it affords the best opportunities and means. When so tempered, parental influence suggests and inspires voluntary submission and obedience. Exercised with love, regard for the tender susceptibilities of babyhood, and sympathy, even denials and restraints lose the sting which impatience and punishment impart to them. The lessons taught by the salutary example and influence of parents who feel the responsibility of the trust involved in the training of children are of the utmost moral value; whereas, when these duties are exercised impatiently, inconsistently, and with petulance, resistance and ill-tempers are provoked. These may be impotent and unobserved in the infant, but will eventually produce the appropriate consequences in the shape of a harvest of baneful fruits of which the sowing-time may have been forgotten. The fault will probably then be referred to those convenient excuses—natural depravity and inherited tendencies.

The neglect of parental duty in the early training of the Will leads to serious consequences. The loss of the providentially appointed time is irretrievable, for when the Will has acquired its strength under the influence of indulgence it is not amenable to gentle suasion, and is exhibited in headstrong wilfulness and resistance to the unaccustomed restraints of authority. Then, as discipline must be observed, it may have to be coerced. Resistance will thus be fortified, temper aroused, and a train of evils and disorders may be evoked, such as is too often exhibited in the spectacle of a spoiled child. Thus, when the Will is developed by indulgence and is unused to discipline, it becomes a dominant power in the mental estate, and contends but too successfully against the authority to which it would, at an earlier stage and by the appointed means, have yielded submissively, and most probably with a loving, spontaneous obedience.

2. The exercise of physical force, or of any other form of coercion, in the treatment of the infant Will provokes resistance in the strong and resentment and vindictive feelings in the weak. It thus begets evil in some of its most malignant and demoralizing forms. It is easy to overcome the feeble powers of an infant, and hence coercion is the usual resource when difficulty or opposition arises. The natural consequences are not considered, as they may not be manifest at the time, and baby is not supposed to notice or remember such things. It prevails at the time in forcing a seeming obedience or preventing the committal of a prohibited act, but it leaves the Will unsubdued, the feelings hurt, and the natural impulse thwarted. These, however, are not feelings or results compatible with the "nurture and admonition" implied by the moral law. Whenever the coercive and repressive force is withdrawn, and the infant powers acquire strength as well as opportunity, the natural consequences will be most probably developed after the causes have been forgotten.

The lower animals show a good example in this respect. Though their rule of life, the law of their lower nature, is force, the brute parent never employs cruel coercion (unless in most rare and exceptional cases) in the treatment of the progeny.

Even amongst those of fierce and predatory races their habit to their young is mild and gentle. Hence the remarkable uniformity of temper amongst the lower animals in their natural state, when they are not subject to the ill-treatment they too often receive from man. Fierce conflicts ensue in after-life, but these are generally between adults under the influence of jealousy or in contentions over prey. The unresisting helplessness of the babe may conceal the feelings he suffers, and the wounds so produced may fester unseen and unsuspected, but they are not therefore harmless. They go towards determining his relation with the world in general, or the temper and disposition of his mind, and they leave the Will unsubdued, though, it may be, for the time overcome.

When the discipline of home-life is affectionately and judiciously exercised, the Divine institution of family confers its greatest blessings and advantages, and fulfils its proper purpose, the young scion proceeding thence into the busy world with the seeds of good habit, with established faith in authority, and a Will under the salutary control of moral principle. But when home discipline has been coercive, productive of mere outward conformity, without the true principle of spontaneous submission, the hopeful-seeming youth, freed from the restraints of his home-life, breaks out abroad into courses which bring shame upon himself and upon the mistaken system which repressed and coerced the Will when it should have been moulded by the unfelt influence of parental love at the proper time.

The effect of forced restraint where the true

spirit of voluntary submission did not exist was terribly exemplified in the cruel and ultra-brutish ferocity of the Indian Mutiny. Nor was there ever recorded in history an instance in which the real character of the Will and disposition were so effectually concealed. No doubt the opportunity and the occasion were exceptional, as was also the scale of their magnitude. Hence the phenomenal character of the proceeding, as it appears on the page of history. Nevertheless, it shows a natural development of feelings, of which minor manifestations occur in individual experience sometimes even in children. We remember a little spoiled boy who vented his helpless rage to an elder brother by promising to "kill him to bits" as soon as he was big enough.

3. Consideration of consequences, as a motive whether of the hopes or the fears thereby inspired, is merely a prudential, not a moral impulse, excepting when it relates to religious ideas. In this latter view it is discussed under the heads 6 and 7—moral principles and Divine ordinance. As a prudential motive, it has little, if any, bearing on our subject. It is a result of experience, and is manifest in the burnt child, who dreads the fire, and both avoids it himself and interposes to rescue a fellow from the suffering of his former experience.

4 and 5. The restraints of human laws and of social obligations supply effective motives of conduct, but are outside of the scope of our argument, as they have no bearing on the education of the Will or its controlling action on the mind.

6 and 7. These considerations are based upon the discrimination between good and evil, and they

determine the relation of the higher and the lower nature in man. They impose on him the conflict by which is established the prevalence of one or the other of those fundamental principles as the actuating motive of conduct. They decide the character and the rule of the individual life. Good is that which conforms with the Divine command and exacts the subordination of the Will thereto. Evil is what conflicts with the Divinely appointed rule and opposes the Will of God.

The appetites and impulses of the lower nature are in themselves good. They are appointed by the Almighty Creator, and constitute the law of the lower animals. They therefore form a characteristic of a large portion of animated creatures and a necessary part of the works of God. They are the means of procuring for man his greatest temporal blessings and his highest sensual enjoyments. They are essential to his welfare in the present life, and are altogether good in so far as they are lawfully used.

The wisdom and beneficence of the moral law are manifest in the effects produced by its observance on the one hand and its violation on the other. In the former case, all the blessings derivable from the lower nature are realized without the penalties imposed by the abuse of its impulses, and, at the same time, these blessings strengthen the power of the higher nature. The subordination of the lower nature to its proper function, as the fulcrum of the higher in the attainment of the conditions necessary for a spiritual life, is the end and aim of the conflict St. Paul describes. On the other hand, the natural operation of lust is manifest in the scenes of degradation and misery which are unhappily too common in large communities. It would be impossible to understand or to conceive of such consequences if they were not thus plainly and practically exhibited. Even with these evidences before men's eyes, the force of uncontrolled passions and appetites often overrules the force of the Divine command, and even the best of men need an Advocate with the Father to plead their cause and aid their efforts. The control of the lower nature affords the special sphere for the exercise of the supreme function of the Will directed by the moral law. The issues of the conflict are spiritual life through Jesus Christ or a second death. The nature and progress of this conflict, which should engage the lifelong energies of the Will, are exhaustively discussed in the pulpits and religious literature of this age, and would not be appropriate to this work, which deals only with the training proper as a preparation for the enterprise.

From the foregoing remarks on the rise, progress, and functions of the Will, it appears—(1) That unless it be governed by moral principles, both in Thought and conduct, it will be under the dominating influence of the lower nature. (2) That its subordination to moral principles cannot be effected by coercive measures. These may *seem* to prevail, but then the outward conduct proceeds from lower motives, and is not the exponent of the heart. The individual is not what he seems to be. (3) That there is an appropriate time for beginning the training of the Will before it has acquired either force or direction of its own, and is virtually in the power of tutors. (4) That the treatment and management of the Will may then be made the means of inspiring ideas of moral obligation, as well as of moulding the Will thereto.

(5) That when the thoughts of the heart, whence proceed the issues of life, are under the control of moral principles, the conditions are those most favourable to the development of the mental faculties, the enjoyment of the life that now is, and for the reception and enjoyment of that spiritual life which inspires the highest hopes of man.

The rise and first direction of the infant's Will having been initiated under the influence of tutors, the further development of its powers will depend, in a gradually increasing ratio, on the suggestions of his own mind, until it eventually devolves on his own impulses entirely. After the child begins to act for himself with a measure of independence, the influence of tutors will be proportionate to the confidence they have inspired. Example will always be in a greater or less degree suggestive, and the environment will necessarily be a chief condition of the mental and moral progress. These influences and the suggestions of his own mind will determine the further growth and direction of the Will. We have therefore to consider what means are best adapted to the training of the Will in order to secure the effective control of the thinking faculty by which moral principles are to be apprehended and applied to the regulation of conduct.

The first step in the process of training the Will is the critical task of giving the first direction to the earliest indications of effort. This involves the highest responsibility, and will require all the tact and patience the maternal mind can exert. At the same time it will afford the amplest scope for the exercise of a mother's love and tenderness. The object is to win a calm acquiescence and voluntary co-operation of the babe, in such manner that he should be insensible of the influence by which he is being led. The manner in which this may be done is not easily reducible to rule, and will be better understood than described. Practically, such power may be exerted even upon adults, whose Wills are, unconsciously to themselves, led by others to an extent which would be incredible—indeed, unintelligible—if it were not actually exhibited in the modern revival of an ancient practice of "*willing.*" The exercise and abuse of such practices were evidently known in Buddha's time, for he denounces them emphatically in his comprehensive code of precepts. Seeing, then, that such power may be, and actually is, vicariously exercised even on the adult mind, we need not contend for the practicability of the natural, legitimate, and effective influence a mother or tutor may thus piously exert in giving direction to the babe's first efforts of volition.

In practice we have often watched with intense interest the delicate measure of loving suasion, so nicely adjusted, so tenderly enforced, and so wisely ordered that the babe under such influence was insensible of any other power than that of a mother's love. Time, patience, tact, and vigilance were not spared in the work of adapting the tension to the tender susceptibilities and feeble powers of the babe, and to soften the strain by all the most familiar and captivating arts of endearment—chiding so sweet as to conceal rebuke in love, firmness so gently swayed as to mould the supple Will by the overwhelming power of *good.* Thus, by the natural exercise of what seems to be a natural impulse of the maternal mind, may the Will of her babe be directed aright, without exciting a sense of denial or provoking resistance or resentment. Nor could any mother fail, if her efforts were stimulated by a due sense of the vital importance of the trust involved in the training of the infant Will. The strongest natural tendencies could not resist such judicious, patient, and persevering effort as we have endeavoured to depict.

Even the headstrong resistance of a spoiled child, whose uncurbed Will had been exerted habitually and had acquired power (that terrible demoralizer), has ceded to the same gentle domination of a wise and loving tutor, whose resources of patience and perseverance were equal to the treatment of the stubbornness of an unrestrained Will. In such a case, resort to force in any form is most destructive of the chances of success, the means most certain to fail, and, in failing, to do the worst mischief. This is not overcoming evil by its proper counter-agent, good, but to provoke resistance and to create the worst disorders of the heart. If the Will be well directed at the first, and so commence its own independent tendencies, the work of tutors in the later periods will be simplified and facilitated, and those adverse influences which must inevitably be encountered in even the purest of this world's homes will be most advantageously met and resisted.

After a right direction has thus been given to the exercise of the Will in infancy, the further training should be pursued in such manner as to suppress and supersede all those tendencies which would weaken control of the thinking faculty or divert the mind from wholesome currents or habits of thought, and, on the other hand, to cultivate such as strengthen the control of the faculties and direct them in accordance with moral principles.

Of the former class we notice—(1) that kind of indifference which enervates the mind and tends to a habit of listlessness, (2) the undue indulgence of appetite, (3) harbouring and brooding over desires suggested by appetite, (4) indolent habits, and (5) habitual dependence.

1. The habit of listlessness described in the last chapter is so influential in impairing the power of control and in fostering injurious habits that it should by all means be discouraged in its first indications. This may best be done by encouraging the active employment of the mind. The habit of suppressing the natural activities and inquisitiveness of children whenever they happen to be inconvenient or to appear frivolous tends to beget a reticent, close habit, and to encourage the indulgence of morbid reflections. On the other hand, sympathy and reasonable encouragement are effective incentives to mental effort, and beget activity and openness. The confidence inspired by the sympathy and encouragement of tutors gives them great influence in directing the natural propensities of pupils.

2. Indulgence strengthens appetite, and simultaneously weakens the power of resistance. It gives both direction and strength to Thought, and the feelings or appetites so indulged become obtrusive, and occupy the mind with unwholesome effect. The influence of indulgence extends over the whole domain of Thought and feeling, and is not confined to any particular exercise of the Will.

Indulgence in the use of stimulants so excites appetite and weakens self-control as to enslave its victims. The same demoralizing effect is produced by the indulgence of other feelings and appetites.

Indulgence of doubt, discontent, or other feelings operates similarly. Unrestrained doubts tend to make the judgment impervious to evidence, in the same way that indulged appetite overrules moral restraint. Discontent likewise deprives even the luxuries of life of their satisfying power, and makes its votaries in the palace more dissatisfied with their wealth than the poor man is with his pittance.

History shows how indulged ambition, unrestrained by superior authority, makes men insensible of the sanctity of human life, and has ruthlessly sacrificed its victims in the secrecy of the dungeon and on the battle-field.

Of all indulgences, that of conscious power unrestrained is the most demoralizing. The good King David, with all his godly exercise and strength of moral principle, fell a victim to the unrestrained power of his regal position, and compassed the death of Uriah to gain possession of his ewe-lamb.

Even priesthoods, in the exercise of their holy functions, when they have forgotten their ministerial office and position, and have usurped the powers that belong only to their Supreme Head, have exhibited to the world abuses of power unsurpassed for cruelty and oppression—as witness the Inquisition and many other of the darkest pages of history.

The only safeguard against the manifold abuses of indulged appetite and passion is to be found in the counteracting influence of good impulses and habits, under the control of a Will conformed to the Divinely inspired prayer, "Thy Will be done."

The Christian doctrine of self-denial is the appropriate antidote and counteracting influence for overcoming the natural tendency to indulgence. Thereby the power of resistance to temptation is fortified. It is therefore one of the cardinal virtues of the Christian system, and one to be sedulously cultivated for the sake of the double influence it exerts in aiding the power of the higher nature and in subduing the forces of the lower.

3. Closely allied to indulgence, in its effect on the control of Thought, is the harbouring of thoughts of indulgence and a habit of brooding over unwholesome currents of Thought. Whether the harboured Thought take the positive form of desire or the more seductive form of controversy with temptation—whether in meditating indulgence or merely of weighing the pros and cons—the end and tendency are the same, and alike inimical to self-control.

The indulgence of thoughts suggested by ill-feelings of all kinds is doubly injurious, for it tends to absorb attention from worthier use, and to occupy it with schemes and projects for attaining the suggested ends. A sense of injury, magnified by frequent reflection and indulgence, will generally lead to ideas of revenge, and thus evil begets evil, and is multiplied. Many a suicide owes the loss of his life and its priceless, irretrievable opportunities to the indulged contemplation of the idea, and the consequent increase of its force and weakening of that control which would have saved him from so miserable a fate.

Reflective tendencies may, however, be turned to good account when directed by habit of control to useful and beneficial currents of Thought. So employed, they may be made the means of cultivating the mental powers, including the all-important one of effective control, which cannot be more effectually strengthened by any other means than by salutary exercise. The facts and laws of science, the beauties of Nature, the blessings of life and even its trials and afflictions, and the dispensations of Providence supply an inexhaustible fund of subject-matter for the useful occupation of the mind and for the development of the reasoning powers. Such use of the faculties is the most effectual preventive of the pernicious habit of brooding over injuries, real or imaginary, or of harbouring sensual and unprofitable thought.

4. An indolent habit is injurious to both physical and mental powers, and gives a general invitation to those ever-present feelings of appetite and passion which are apt to make themselves so obtrusive. Evil companions, whether those of Thought or of other idlers, are dangerous tempters. The Will that consents to the waste or loss of *time* (the stuff that life is made of) must have attained a serious demoralization. Idle minds, like idle hands, are generally turned to ill use.

Industry of body and mind strengthens the faculties, and is in itself a form of habitual control. A life of earnest occupation in useful pursuits runs its course more smoothly than one of profitless indolence, has fewer vexations, more satisfaction, and greater reward than one of luxurious ease, so called. It earns the commendation of a good stewardship, and escapes, besides the terrible condemnation of the buried talent, the unspeakable humiliation of self-reproach. In that dread hour when death is at hand, and the last sands of life are fast running out—when reflections on the past life force themselves on the mind and suggest the verdict of the great Judge—a sense of its uselessness must be most humiliating, even if the vices which idleness almost always induces had not also to be deplored.

5. A habit of following a lead, and depending on the opinions and suggestions of others, in so far as it is a voluntary renunciation of the active control of Thought and conduct, is one of danger. The weakness of principle and infirmity of judgment implied by such a habit are inimical to effective control. He who, for want of self-reliance, seeks for a guide will most probably fare like the unclean spirit, and find others weaker than himself to occupy his empty mind. The control and the responsibility of conduct cannot be safely disconnected, nor may they be deputed to another.

That form of distrust of self which is a characteristic of the conflict between the suggestions of the lower and the principles of the higher nature, which drives its subject to the Source of unerring Wisdom and Power, is far different from the desultory dependence of an infirm mind. Then, as St. Paul says, when most sensible of his own weakness, he finds the strength he requires, not in a wandering, uncertain search, but by direct application to the proper Source, the Fountain of all strength and virtue.

Such scenes as those exhibited by mesmerists, of which the scientific explanation was given by a Dr. Baird, of Manchester, over forty years ago, show the effect of certain forms of renunciation of the individual Will. Dr. Baird published the result of his researches on these phenomena under the name of Hypnotism. He frequently lectured on the subject, and exhibited on his platform the effects induced by his methods on numerous persons, comprising volunteers from his audience and subjects introduced by himself. Complicity and fraud were altogether out of the question, several of the subjects being well-known residents, some of whom submitted to his treatment incredulously, but all who followed his directions fell under his influence, and showed the most abject obedience to, and dependence upon, his Will. In all such exhibitions, as also in the more modern forms of submission to the Will of others in "willing," the danger of trifling with and abusing this governing power of the mental estate is plainly manifested. The practical lesson so taught is to show that that power should never be deputed. Man may not, in fact, with impunity trifle with that power which he is bound by Divine command to mould to conformity with the Will of God.

The cases of possession mentioned in Scripture may probably have had some affinity to these strange phenomena. In some forms of nervous disorder the power of the Will seems to be impaired by want of salutary exertion and use. A bedridden invalid who could not be aroused to the use and control of his faculties, though his doctor was convinced of his physical ability to do so, sprang out of bed on a sudden alarm of fire being raised. His Will, under this strong stimulus, was effectually exerted.

The practices denounced by Buddha were evidently of the nature of those above mentioned, wherein the individual Will is dethroned and superseded, and his denunciation of them proves the demoralizing effect so produced. All these allied phenomena have an important bearing on the subject of this chapter, as they all depend upon unhealthy action, temporary suspension, or permanent disorder of the motive-power of the mental estate. All alike demonstrate the danger of neglecting the natural and salutary exercise of this principal faculty, and attest the terrible results which attend its dethronement, however temporary that may be.

No other subject can exceed in interest or importance the control of that faculty which determines the character in the sight of God, which furnishes the equipment and measures the fitness of man for his functions in this world and for his enjoyment of that which is to come—which affects his faith even more than his feelings, and establishes his relations with his Maker even more than his connection with his fellow-men. Thought is always present, even when external circumstances are absent or powerless, as in the hours of sleeplessness or other abstraction; and they determine the brightness or the gloom of the present, animate the hopes or inspire despair of the future, suggest and stimulate good works or project evil, enkindle love or nourish hatred, use the faculty wisely or waste it wantonly, according as the currents of Thought are directed this way or that, to entertain and cherish the good or to permit and follow the evil.

Hence, the choice of the thoughts determines to a very great degree the sweetness or bitterness of their companionship, the happiness or misery of life, and the contentment or anxiety of mind. The power of controlling Thought confers the ability to pursue with effect any particular currents or course of reasoning and to discard in their favour every disturbing influence from within or from without. On this power, too, depend uninterrupted worship of, and communion with, God, free from the distractions of external circumstance or internal suggestion. There is no other position in life in which the want of the controlling power of the mind is so distressing as in intercourse with God in prayer. Interruptions are then destructive of peace and profit. The incongruity of suggested Thought at such times is often simply appalling.

The power of choosing the currents of Thought and directing the powers of the thinking faculty is vested in every individual for himself. This proposition, so vitally affecting the issues of life, may seem doubtful to persons who, for want of training, or owing to the effect of loose habits of thinking, have lost, or have failed to acquire, command over their faculties. It is nevertheless assuredly true that every sane man is endowed with a power of Will which, rightly exercised, gives him the mastery of his mental faculties as complete and general as of his physical powers. He may even learn to overcome the pressure of external circumstance, and, discarding all else, keep his mental powers uninterruptedly concentrated upon the object of his choice. Lord Macaulay relates that he could keep his attention fast rivetted upon his book as he walked along the thoroughfares of London. His may have been a singular supremacy of control, but the accomplishment of withdrawing the attention completely from surrounding objects is one which is required in many of the pursuits of life, and though it may sometimes cause difficulty, and be imperfectly acquired in some cases, these must be exceptional. In schools, offices, and many other places in which daily avocations are pursued, the individual efforts have to be independent of the concerns of absorbing interest going on around.

It may therefore be confidently affirmed that every man either possesses, or may acquire, the control of his thinking faculty in such manner that he may choose out of his fund of collective ideas those he will entertain, commune with, and cherish, and may effectually discard others which would interrupt or divert his attention. Hence it follows that the functions and influence of Thought on life and character being such as we have described, and every man possessing the means of exerting effective control of his thoughts, the tenor of life in every individual depends mainly upon his own action in over-ruling the force of external circumstances. It rests practically with himself whether he persistently direct his thoughts to the dark or the bright side of things; whether he brood over his disappointments, and thus multiply them, or dwell upon the blessings of his lot, and draw from them the full measure of solid satisfaction they would never fail to afford; whether he nourish resentments and cultivate the company of the injuries he thinks he has suffered, or whether he obliterate all such miserable memories by the cloak of charity. He who chooses to compare his lot with that of others whom he deems more fortunate than himself makes himself a discontented man, whilst he who, on the other hand, chooses to contemplate the blessings he enjoys creates a contented mind, and enjoys them all over and over again. He makes himself happy with a fraction of what the other permits to make him miserable. Jack's horse is a familiar case in point. He never regarded his own pasture, but spent his time in looking over the fence into the next field. Thus it is manifest how greatly the happiness or misery of life depends upon the control of the thinking faculty, and therefore rests with the individual himself.

It is true that painful subject-matter may sometimes obtrude itself upon the mind with irresistible force—such as the failure of some cherished project or the miscarriage of some earnest effort. Or the loss of some dear relative or friend may inflict a wound from which the heart must suffer and the current of Thought draw copious grief; or affliction may assail, and impose its inevitable pains and penalties. Still, these have a mysterious and salutary influence, unless they be permitted to do more than their appointed part in the economy of life by uncontrolled licence of Thought. Many others since the days of the Psalmist have found it good to have been afflicted. All such occasions supply the sphere proper to the enlistment of sympathy, the great alleviator. St. Paul speaks of such affliction as *light* in view of the eternal weight of glory which they are in some measure the means of procuring. These are all opportunities for the higher nature to assert its supremacy over the affairs of the lower. As such they are essential parts of the dispensations of Providence for the exercise of faith and for the education this life affords.

Nor should recollections of sins be resisted. The grief they inspire is an essential condition of forgiveness and of that genuine, heartfelt repentance which drives the culprit to the Cross. Moreover, the remembrance of sin is a most effective stimulus to the love of Him who expiated the sins of mankind by His exemplary life and sacrificial death. The love of the Magdalene, which so excited the self-sufficient Pharisee, was proportioned to, and inspired by, the recollection of the debt her Saviour had discharged for her.

In all those cases the direction of the thoughts to subjects in themselves painful does not produce remorse—that hell-fire of the heart which is the sting of the second death. Unless unduly indulged, the effect of such contemplation is that of a cleansing fire, a purifying, chastening influence; it is therefore not the least profitable company a man may sometimes select from the treasury of things new and old.

He who, possessing a palace, should take up his abode on a dunghill would be accounted mad. Yet he who, having a treasury of life's experiences, should choose for his meditations all the bitter elements thereof, and should neglect to consider the sweets, which in every life, even in the most afflicted, far outweigh the real bitters, acts likewise, and with no less unwisdom and bad taste. However widely the social position and circumstances differ, there is in every life ample material, in the experiences of good and evil, from which to make choice of companionship in Thought. No sphere can claim any monopoly of choice. All depends on character, which is the exponent of the thinking faculty, and its product, Thought—on what has been extracted from the opportunities of life, either under the salutary control of moral principles or under the captivating influence of the lower nature. This it is which furnishes the mental treasury, and therefore, in whatsoever sphere of life or grade of society his experiences may have been acquired, every man has a variety of resource from which he may choose either the good or the evil, according as his wisdom or his ill-humour may dictate.

Considering how large a proportion of life is employed in reflection when no other resource is available, it behoves every one who values happiness and peace to make provision for the employment of his leisure thought such as most men make for their hours of physical recreation. Most men have places of resort where they enjoy their pastime in company with the beauties of Nature or the recreations of art. They avoid the places where vice and folly disclose their revolting characteristics. They prefer the fragrance and loveliness of the flowers in their gardens to the drains and dunghills appertaining. So, likewise, mental pastime may either have its pleasurable and useful resorts amongst the attractions of scientific, literary, or religious Thought, or in unwholesome broodings over those things which have embittered their past lives and may prove the hell-fire of their future state. Every man should possess, as an essential part of his mental equipment, a repository of reserve Thought from which to draw in time of need such matter for reflection as may best suit the particular occasion—a mental pleasure-ground in which to take his mental pastime. Thence, if sleep delay its sweet solace, as he lies in the stillness of night, or if in a vacant moment dark Thought assail, or in any of the innumerable occasions of mental leisure, he may always derive an unfailing supply of salutary or useful and agreeable Thought to occupy the spare moment or to repel an obtrusive suggestion. No man's experience of life is so barren of happy reminiscences that he needs brood over the incomparably fewer incidents of sad memory. The kindness of friends and the pleasant intercourse by which they have so often sweetened life, the love that has so often been returned from sympathetic hearts, the joyous recollection of good deeds done and the still more delightful projects of others in contemplation, the beauties of Nature, the triumphs of art, the gifts of Providence, the goodness of God, and other cognate reflections are ample defence against the assaults of the lower nature with such poor baits as thoughts of injuries suffered or other sad reproductions and multiplications of misery.

In conclusion, the control necessary to firmly establish the habit of choosing wisely the subject-matter of Thought is an attainable accomplishment within the reach of every one. And its attainment is an end worthy of the most sedulous and persistent effort from the first exercise of the mind in infancy to the very end of life. And it would still be so even if it were not the duty of man to eschew evil and choose the good equally in the domain of Thought as in the conduct of life—if, that is to say, it were not as imperative to cleanse the fountain as to purify the stream.

In this chapter we have endeavoured to show that all mental operations are naturally subject to the Will, the governing power of the mental estate; that the effective maintenance of the supremacy of this control depends upon its being regularly exercised, whereby it becomes habitual, easy, and almost mechanical; that, on the other hand, if not so controlled, the mental faculties will fall under such other influence of internal suggestion or external circumstance as may happen to present itself, and that the more this *laissez-faire* habit is indulged the stronger it will become. In the end, control will become impossible except by the exertion of extraordinary effort. The habit of listlessness, therefore, is the natural counter-agent of control and of attention, its exponent. To " collect " Thought which is habitually dissipated is but barely possible, and is always a painful and humiliating task, like tearing off the shackles of a mental bondage.

The practical conclusion is that, from the first, mental effort should be fixed upon whatever matter is actually in hand. That is its natural function. It should be engaged in whatsoever occupation is present. Leisure is always a snare, especially to children, who have but little mental resource to engage their reflections. It follows that employment, of whatever kind, not positively harmful, that possesses interest and attracts attention, is doing good by keeping Thought enlisted to a purpose, fixed on a pursuit, and not left to drift into such currents as are ever ready to engross idle minds. When the Will renounces its prerogative, its power wanes, and leaves the mind a prey to desultory Thought and a habit of listlessness.

The most effective way to cultivate a habit of attention in children is to encourage it by entering into their pursuits and by sympathizing with their own natural efforts, and thus helping them to keep their thoughts engaged in what they are about. But little ingenuity is needful to devise attractive employments for young children, because almost any may be so made by sympathetic attention of their tutors.

CHAPTER XIII.

ON MEMORY.

Without memory there could be neither past nor future. As there could be no recollection of the past, there could be no conception of a future, for this is the result of reflection, and without memory there could be no reflection nor any ideas to reflect upon. The mind could not possess any stock of Thought, and could not have any knowledge except of the sensation present, which, when past, would vanish out of knowledge and leave no trace behind. Each sensation would be a separate and detached bit of life, unconnected with all others. There could therefore be no individuality, for life itself would consist of disjointed links without any connection to give them either unity or continuity. Thought would be but the shadow of the passing sensation, and would disappear along with it. The mental tablet would be as blank at the close of life as it was at its commencement.

It is not altogether unprofitable to consider what life would be without memory, because it helps to show the inestimable value of that faculty as being the one on which all other mental powers depend. This could in no other way be so forcibly exhibited as by the endeavour to conceive what life and mind would be reduced to if deprived of memory. Even the lower animals supply no clue to aid the conception, for wheresoever mental power exists, even in their lowest ranks, it is shown in memory.

Practically, the nature of memory is so well known by universal experience that no words could make it better understood than it already must be to every one capable of understanding the words employed in describing it. Nevertheless, the essential nature of the faculty and its mode of operation are inscrutable. All that is, or seems likely to be, known respecting these points is the practical operation of the faculty. Respecting this there is no mystery; nothing, in fact, but what may be well understood by experience and research. And there is no other faculty which merits such attentive study, for its power is the measure of all other mental powers. Even the perceptions derived directly through the senses owe their chief value to the relations subsisting between them and existing experiences of the mind. For all the qualities of things—such as their size, distance, and the like—and their uses and properties are learned by means of comparison and inference, for which memory is essential.

The records of formulated Thought, the ideas inscribed on the mental tablet, are of three kinds—simple, complex, and indeterminate. Simple ideas, like single letters of the alphabet, are rarely apprehended or employed singly, but as component parts of complex ideas, of which class are the great proportion of the mental stock. Both the first-named classes of ideas are definite, and are therefore recallable in definite form by the memory. Those of the third class are not so, but, though indeterminate, they are nevertheless often deep and enduring.

They may generally be expressed or described by means of other experiences, but they are not always translatable into customary forms of expression. Of this class are those impressions or ideas formed of sermons, narratives, and books, of which no particulars were separately noted, and of which only a general idea is remembered, albeit this may be strong and permanent. These several kinds of ideas are fully described in chapter ii., and we have now to discuss other characteristics of Thought not there specified.

Ideas are not only of different classes, but those of the same class may differ widely in strength, clearness, and other characteristics, according to personal idiosyncrasies, such as the habit of mind, manner of observation, and effort of formulation.

Habit of mind may, for example, be energetic and earnest or supine and indifferent, or it may be careful and attentive or casual and desultory. These and other personal characteristics affect the nature of the ideas formed under their influence. Other differences arise from the various habits of observation which are peculiar to different individuals. One regards details, another takes more general and comprehensive views. One attends to the sense of a phrase more than to the form in which it is expressed, which will engage the notice of another preferentially to the sense. One is impressed with the separate features of a face, and notes the colour of the eyes, the form of the nose, &c. Another, neglecting these details, notes the general expression. His idea is of the *tout ensemble.* One studies the facts of a science individually, and forms imperfect ideas of its general principles. Another regards the facts chiefly in their relation to principles, of which, therefore, he has clear ideas. Other differences in the nature of ideas arise from the act of formulation. One person is more affected by what he *sees*, another by what he *hears*. The ideas formulated by each will differ accordingly. In one the evidence addressed to the eye will be relatively stronger and more complete than that apprehended by the weaker or less used sense, and *vice versâ*.

Hence the collective stock of ideas in each individual depends upon the various conditions and experiences proper to each, and the ideas differ accordingly. In one the stock will consist of a preponderance of details and particulars; in another, mainly of general characters and principles and of more comprehensive views. One person has clear, definite, and complete ideas, which in another are loose, cursory, and partial, and so on *ad infinitum*. It follows that there exists great diversity of memory, corresponding to the kind of knowledge in possession, and to the nature of the ideas of which the stock is composed. For it is manifest that no other ideas could be recalled by the memory than those existing in the mind, and, seeing that all the ideas there present were formed and collected by the individual himself, it follows that the diversity of memory in different individuals is owing to the different kinds and qualities of the ideas which compose the stock in each case.

Almost every individual memory is characterized by special power or weakness in certain directions other than those peculiarities which arise from the greater or less prevalence of particular kinds of knowledge. The origin of these specialities is sometimes obscure, but it may generally be traced to natural causes by careful reference to past experience. For instance, unmeaning gibberish picked up at school may often be remembered through life, or it may be incidentally revived, after a lapse of forty or fifty years, in all its absurdity, as fresh as if it had been repeated daily. Yet, on the other hand, useful and intelligible formulæ, replete with meaning and acquired at the same time by much labour, may have passed completely and irrecoverably out of the mind. The memory which recalls the one fails with regard to the other, though both were originally contemporaneous, and one was picked up casually and the other by laborious effort. The one, therefore, which should have been enduring is lost, whilst the other, of a fugitive nature, is retained. Still, it would be unreasonable and unphilosophical to attribute the difference to chance. It must be due to the operation of natural causes, which, therefore, should be sought for the purpose of turning them to account.

In the particular instance just cited, the idea so faithfully recollected was acquired *voluntarily*, under the influence of imitation and emulation, much in the same manner as that by which language comes with so little apparent effort and with such lasting effect. Regarded as one of those accomplishments which children are often so ambitious to acquire and so proud to display, it was probably the result of oft repeated trials, and practised continually for a considerable time afterwards. Hence it came by voluntary and earnest effort and was confirmed by after-use, and these are the conditions most favourable to effective formulation. On the other hand, the useful and important lessons which have since been so completely lost to memory, were originally *tasks*, not willingly taken up, but imposed by school discipline. They were not coveted accomplishments, but dry and unattractive impositions. They excited no interest, nor seemed of any use. In short, the ideas of them were formulated under conditions the most adverse to the production of deep and lasting impressions, and were therefore, from their very inception, doomed to be forgotten unless soon revived by the future requirements of some definite pursuit. Thus considered, the fact in the case cited proves the fidelity of the memory to the influence of natural causes.

A careful consideration of past experience will, we think, convince any person that the great proportion of what he remembers the best, and of his most useful and available knowledge, was gathered by voluntary impulse from passing events and observation, because it presented some form of attraction to excite desire. It may have promised some present gratification or some future use, or it may have excited emulation or have been prompted by the imitative propensity so strong in children. But whatever the motive, the act of acquiring the particular information or possession was voluntary, self-chosen, and performed with *intent*, and, therefore, with earnestness and effect. These are conditions favourable to the formulation of deep and lasting ideas. Another condition which is very influential in childhood is the absence of pre-occupation.

Another kind of seeming discrepancy in the function of memory is the subject of a very common form of complaint. A busy merchant, it may be, inveighs against his weak and treacherous memory because he forgets sometimes things which he wished to remember. He assumes that, because he wishes to remember, his memory ought to obey his wishes and never fail in such matters. The fact in his case is probably that his power of memory is taxed or perhaps strained to the utmost to remember the vast mass of detail necessary for the exercise of his calling. He fails to consider that every vessel, even the human mind, has a certain capacity, and that when it is full it will hold no more. The last occasion on which this complaint was expressed to us it was in vehement invective by a professional man, of whom we knew that his splendid power of memory was an unfailing resource in all those uses connected with his calling for which it was most needed. A mind so engrossed with matters of paramount importance has not much power to spare for superfluous use, even if its pre-occupation were not itself an effectual impediment to such exercise.

Memory, being the power of recalling ideas previously formulated, will necessarily reproduce every idea in the particular form in which it exists in the mind. This, as we have shown, will depend on the condition of the mind at the time of the presentment, on the manner of observation or contemplation, the habit of formulation, and other circumstances to be hereafter specified. Such as is the nature of the general stock of ideas collectively, and of each particular idea individually, such will be the kind of memory in its general character and in its particular application. Hence, in infancy, for example, when strong impressions are made before any knowledge of language has been acquired, it is manifestly impossible that such impressions should be formulated in words. Nor in that inexperienced condition of the mind could ideas be formed in any other *communicable* terms. But they are not therefore lost. The forces which acted so powerfully were not annihilated. Though the ideas do not exist in any of those impossible forms in which the infant mind did not yet possess the means of embodying them, they nevertheless exist in the full force and form in which they were experienced. These ideas therefore are not recallable in any of those other forms, seeing that they never so existed in the mind. They are crystallized into the form of temper, &c., and they determine the feeling or relation of the infant towards external persons, generally or individually, as the case may be.

The sensation of a burn, for instance, could not be formulated by a young infant either into words or any other terms by which it could be expressed. Nevertheless, the sensation will be remembered, and will result in a manifest attitude of repugnance towards the producing cause of the sensation. In like manner, unkind treatment will produce lasting impressions corresponding thereto, which will be manifest in the form of temper, disposition, and feelings of fear, resentment, or dislike on the one hand, and their opposites, love, trust, and allied feelings, on the other hand. The stock of ideas acquired in early infancy consists, to a great extent, of such indeterminate impressions, which even in after-life are not always easily expressible. Nevertheless, the experience of that period is influential in the formation of the future character, inasmuch as it determines the relation of the infant to the outer world.

In later periods of childhood, when knowledge is still crude and language is imperfectly understood, ideas formulated in terms of such defective means partake of the imperfection of the means employed. The idea of the house in which a person passed his childhood is clearly and distinctly embalmed in memory, but on revisiting it twenty years later he will probably be astonished to see how unaccountably it has shrunk in its dimensions, how changed are its aspect and surroundings! The dame-school of which he has such vivid recollection will be found to be incredibly transformed, and the great ogress who used to slap his little hands seems to his mature senses a nice little old woman, all smiles and curtsies. She and her house of correction, depicted in the colours and dimensions of his childish standards, now appear in another light, and are judged by other standards, corrected and established by long experience. We could hardly believe our senses on being introduced, after a lapse of twenty years, to the real presence of our first schoolmaster. Till that disillusion occurred he had lived in our memory in figure clear as day and dark as night. A shudder involuntarily shook our frame when the image of his spectacled face and portentous figure crossed our mental path. We saw him examining our blotted copy and chastising our inky fingers. The awe inspired by his comments on the crookedness of our attempts at straight strokes still haunted us when in memory he re-appeared to us. Yet *now*, how different does he seem! It seems impossible that this courteous, kind, polite old gentleman can be the very original of pictures so unlike. But so it is, and the change is not in the original. It is the copy that was imperfectly drawn, on a scale which experience has changed, and in aspects referred to standards long since superseded by others, truer and better, from which a part at least of the original imperfections have been supplied or corrected.

From such considerations it appears that ideas are formulated and chronicled in the terms, and by means of the standards, proper to the period when they were produced. The size, colours, and aspects of objects without, and of subject-matter of Thought within, are all gauged according to the standards by which they were judged at the time when the ideas of them were formulated and inscribed on the mental tablet. Ideas of external objects, as well as those of personal treatment, were true to the mental standards of the time when they were formed, but, whereas those have not affected conduct, these, magnified by the sensitiveness of infant feelings, have deeply affected the child's subsequent relations with the world.

All the imperfections and deviations from truth that exist in the standards by which sensations and reflections are judged at the time they are formulated are reproduced in the images and ideas resulting therefrom. The effects in regard to such matters as are described in the foregoing paragraph are evident. But though the effects are not so obvious, they are equally certain and infinitely more important in their bearing on the processes of Thought in the later periods of life. These are in fact distorted, deformed, and imperfect in like manner and degree owing to the like infirmities. The formulative faculty which inscribes ideas in the mind, the terms in which the inscriptions are recorded, and the nature of the records themselves are all imperfect, and more or less erroneous in sympathy with the various forms and degrees of infirmity proper to each individual thinker. Hence the records habitually summoned by memory for use in after-life, and all the opinions and judgments founded thereon and the conduct suggested thereby, are all tainted with the ineradicable elements of error which exist in all human standards.

Considering that the ideas which compose the collective stock in every case are the data, and the only data on which all judgments and opinions are formed, and that opinions so formed on previous data become in turn the factors of further inferences and judgments *ad infinitum*, it is evident that error and imperfection in various degrees are inextricably mixed up with all, even the wisest and best of human reasonings. This fact is humiliating truly, but it is irresistible, and should teach men to be tolerant towards those of their fellow-men who, having formed their opinions under different conditions, think and act differently from themselves. And further, it suggests the utmost care in admitting into the mind ideas from which error has not been, as far as possible, excluded, and also in forming opinions on any but the most reliable premisses.

The memory is the index of the mind, and faithfully indicates the thoughts and experiences that are recorded there. These are reproduced by memory in the form in which they stand there recorded, whether in their original garb, as they were first formulated, or as they may have since been revised and corrected by subsequent repetitions. As they stand on the record at the time of memory's summons, so will they re-appear. They are available for use in that form, and in that form only. He who has collected details only may in vain expect to recall ideas of general principles. The listener whose attention during a discourse was fixed on the line of the argument need not hope to exhume from his mind the beautiful phrases and periods of its diction. Nor will he be much profited by a sermon who is chiefly occupied with other thoughts. He may tax his memory, but he will not find there anything but what was attentively received or very strikingly enforced.

Listening once to a celebrated orator, we heard his opening remarks with such attention that we could afterwards repeat them word for word. But, as the discourse proceeded, the subject-matter became more attractive than the diction, and absorbed our attention. The consequence was that from the time when the argument occupied our mind the diction was lost. Not a phrase of it could be recalled. But this was no fault of the memory. The reason we could not recall the words was because we had not formulated them. They were not on the record.

A witness in the box is required to give his evidence, and we will suppose that he has a strong sense of the obligations imposed by his oath and of the importance of his testimony. He remembers the incidents he is required to attest quite distinctly, but the events were observed and formulated by him at the time of their occurrence in a certain way. In the box he is required to answer certain questions in a certain form. But probably his record does not supply a ready answer, if any, in that particular form. He hesitates possibly, if he be not altogether outfaced by the gazing multitude and the court; he proffers an explanation, but is promptly stopped and probably bullied, as though he were trying to conceal something or to prevaricate. He is bound by the terms of his record. His memory will not accommodate itself to the form in which the counsel is dealing with the facts, and he therefore appears to be at fault. If allowed to tell his story as it stands in his mind, he would probably give it clearly.

Simple and undeniable as the fact is, it is not always recognized in practice, that memory cannot reproduce the ideas of the mind otherwise than as they stand recorded. *Memory* cannot alter the record; that is the work of *fiction*.

Hence, the fact that each individual has a particular habit of mind, a certain manner of observation, and formulates his ideas in his own way, determines the character of his memory, which is only the index of what the mind contains. So that whatever may be the nature of the ideas in possession, and in whatever form they exist, such both in form and substance will be their reproduction by the memory—general or particular, clear or obscure, true or false, complete or imperfect; as they exist in the mind, so they will answer to memory's summons. Yet the chief complaints that men make of their memories are because of their fidelity to the record. For example, a person of careless habit is annoyed to find that his recollections of certain things have proved to be very faulty, and he blames his bad memory because it has recalled but too truly the ideas he formulated so carelessly and hastily. Another complains how little he remembers of the last book he read, and deplores his wretched memory. He forgets how hastily he skimmed it through, how greedily he bolted page after page, without afforded himself time to fully grasp any but the most salient parts, and even these were passed over so hurriedly and so piled one upon another that the idea formulated of the whole was a disorderly hotch-potch. The memory answers to the performance, which was cursory, superficial, imperfect, and jumbled.

In the reading of a book, or in any other effort of the mind, it is necessary for every one to gauge his own proper powers by experience, and to impose upon himself only so much at a time as he knows he can digest. It is quite as essential to mental health and strength to ascertain the capacity of the mind, and to adjust mental food accordingly, as it is to control the measure of the meals, and adapt them to the capacity and digestive powers of the stomach.

The circumstances and conditions which determine the *form* of the ideas which compose the mental store have been already sufficiently discussed. But, although the *form* they possess in the mind is that in which memory will reproduce them, the facility with which they may be recalled depends on other characters equally important. The readiness of recall is often even more important than the form of an idea. The conditions necessary to insure a ready response to the summons of memory, therefore, require special consideration. In other words, the *force* or strength of an idea is as important as its *form* in relation to the action of memory.

The conditions which chiefly determine the *force* of ideas, and their consequent readiness of recall, are: (1) the attention or effort of concentration exerted in their formulation, (2) the nature and duration of the opportunity afforded for the act of formulation,

(3) the subject-matter or sensation itself, (4) the condition of the mind at that time, (5) associated circumstances, and (6) repetitions.

Though we do not profess that there is any real relation between the results of photographic and mental processes, there are points of analogy which serve for illustration. Hence, as the actinic strength of the light, the sensitiveness of the plate, and the duration of exposure are all essential factors in determining the result, so, in the action of the formulating faculty of the mind, there are the corresponding elements of force and concentration of attention, the sensitiveness of the perceptive power, and the duration of the process, all effective factors in determining the result.

1. Without a strong effort of attention, either voluntarily inspired by earnest desire or constrained by pressure of circumstance, no effective result is possible. Otherwise, the impression will be more or less light and fugitive. It has already been shown that the act of formulating an idea is distinct from the mere perception of a sensation. It follows that the strength of the idea in the mind will depend on the effort employed in that act. Of all the things seen and heard, a part only enter into the stock of formulated Thought. Of the rest, some things are altogether unnoticed and lost; others are very feebly formulated, and produce weak and fugitive results. The choice of those that are noticed depends in part upon their attractiveness or the obtrusive nature of their presentment, and in part on the condition of the mind and other circumstances hereafter to be mentioned. It follows that ideas of every degree of force exist in the mind according to the degree of

attention devoted to the act of formulation or to its frequent repetition.

An instance of the total loss of a perception the existence of which was proved by responsive but unconscious action is given in the Appendix, Note 7. In this case a person is addressed formally, and a letter is presented to him, which he takes and disposes of unconsciously in so far as the act of formulation or record is concerned. He is asked for the letter, and denies any knowledge of it or of any presentation of it to him. The case represents a common form of occurrence. The letter was delivered to one who was wholly absorbed in an important calculation, and was received at a time when the mind could not, and did not, *attend* to it. The act elicited a mechanical response, and must, therefore, have been *perceived*, but was not formulated, and hence not subject to recall, as no record existed in the mind.

It is not uncommon in the act of reading, even when reading aloud, for whole sentences or even pages to be well read with seeming intelligence, and yet to be utterly lost, without so much as a glimmer of either words or sense. Sometimes the last *sounds* may still linger on the sensorium and be recovered by immediate effort, owing to the fact that sensations linger for a sensible space of time on the sensorium. It is by means of this phenomenon that the curious effect of the zoetrope and some analogous illusions are produced. We remember a nervous youth whose duty it was to attend upon a pompous, irascible principal of a large house of business, and who had a habit of pausing and pondering for a few moments whenever he left the "sweating-room," as the chief's office was commonly called. Partly owing to his pomposity, and partly owing to a bad utterance, the chief blurted out his orders in a most confused manner, and few people ventured to ask a repetition. Least of all the nervous little attendant, whose habit therefore was to ponder over the sounds whilst they yet lingered on his ear, and try to spell out their sense. Manifestly, however, this can only apply to the few last utterances. Hence the effort to follow a discourse so uttered would be unavailing, and, inasmuch as one sentence would follow before the previous one had been apprehended, the whole would defy subsequent recall by memory, for the all-sufficient reason that the resultant impression was a conglomeration, of which possibly, here and there, a strongly accentuated phrase might have been fixed on the photographic plate in recognizable shape.

The force of ideas and their amenableness to memory depend therefore very much upon the degree of attention exerted on the act of formulation. Ideas will be light, shadowy, and imperfect, or strong, clear, and complete, responsive to recall or hard to remember, according as this process has been efficiently or imperfectly performed. It is chiefly to this factor that mnemonic systems direct their rules. The directions they prescribe aim to give special force to the ideas by particular attention to the manner of their formulation. The necessity for such effort of attention is practically recognized. An architect who inspects a site for a contemplated building is not satisfied with a cursory look, or even with bare measurement. He notes its exact figure, the contour of its surface, the nature of its soil, aspect, situation, the views it commands and the view it will present.

He considers it with reference to drainage, elevation, and other points. And he does not expect, even after all, that his memory will supply any particulars he has neglected to note or to correct any mistakes of his measures or memoranda. Yet this, in effect, is what very many people expect their memories to do, and, because they do not answer this expectation, their memories are denounced as treacherous and weak, and such persons often regard themselves as victims of a providential arrangement by which they have failed to receive their due proportion of this most useful faculty.

2. The nature and duration of the opportunity afforded for the contemplation of a sensation, and for the act of formulating it, are cogent factors in determining the result. It stands to reason that, if the sensation be brief and transient, the conditions will be adverse to effective formulation. Or, if the sensation be striking, attractive, and sufficiently prolonged to afford good opportunity for contemplation, the result will be proportionate to the attention given. The nature of the presentment and the opportunity may, however, be either improved or lost according as they are used, neglected, or abused. A very common form of abuse is hasty reading, to gratify present craving or curiosity, or under the influence of an erroneous estimate of the reader's power of mental digestion. A man who reads a three-volume novel as fast as he can scan it through may probably not care to remember much about it, and he would perhaps suffer no loss if he forgot it all. Still, such readers generally attribute their want of recollection to infirmity of memory, and their deficiency of memory to want of some natural power.

The fact is, however, that there are limits of capacity to even the most powerful minds beyond which further effort is unavailing for its purpose, and causes injurious strain. Beyond these limits effective formulation is impossible. Whatever the natural capacity may be, the mind cannot receive more than a certain quantity at once under the most favourable circumstances. This limit of capacity should be ascertained by each individual for himself by means of experience, and it should afterwards be observed by rule. For it is certain that effort exerted beyond that limit will cause strain and injury. There is also a limit of time, or rate of apprehension, which cannot be exceeded with benefit nor attempted without mischief.

The habits of reading more at one time than can be digested, and of reading at a rate exceeding the power of apprehending the sense effectively, have a dissipating and weakening effect on the memory, because they set up a bad habit of reading and thinking which operates generally. The habitual novel-reader, in acquiring a habit of rapid, cursory reading for his present gratification, disqualifies himself thereby to some extent for the effective reading of solid matter for after-retention. Such habits of reading and thinking and of neglecting to formulate and fix ideas in the mind operate adversely whenever formulative effort becomes necessary for acquiring useful knowledge, and they form serious impediments to real learning.

The foregoing remarks apply also to attempts to listen to a greater quantity of matter at one time, or to follow a discourse uttered at a greater rate than the power of apprehension and capacity are equal to. A powerful preacher will often pour forth a voluble discourse of an hour's duration without reflecting that it contains a mass of matter which exceeds the capabilities of nine out of ten of his hearers to digest at one meal. And he utters it at a speed which still fewer can effectively follow. Whilst the most attentive of his hearers are trying to fix his last sentence he has already finished another. Even those who may have had their minds gratified for the moment carry away at best a confused and very indeterminate idea of the sermon as a whole. As to the rest of the hearers, they go away with a sense of pious satisfaction that they have done a duty. Both preachers and teachers would do well to consider the digestive powers of those whom they instruct, and to measure the mental meals they administer to suit those powers. The effect would be to save themselves a vast amount of superfluous labour, and a great deal of futile effort to the recipients of their bounty.

The mental gluttony to be witnessed in exhibitions of art—of pictures, for example—would be amusing if they were not so sad. A man will vouchsafe a look of ten or twenty seconds to each of a score or more of the works of Raffaele, Angelo, Correggio, Guido, and such masters, in quick succession, and, having *done* the gallery, will leave it with the idea that he has fully appreciated and understood those great works, and will offer his criticisms and argue his points with the first artist he meets without perceiving how much he has over-estimated the powers of his mental digestion.

3. The *nature* of the subject-matter or sensation in question is necessarily a very important factor in determining the force and prominence of the idea as compared with other ideas in the mind. The degree of in-

terest a subject possesses will vary very much in different individuals, but, whatever the subject may be, the interest it excites will go far to determine the force of the idea it will suggest. Hence the great importance of inspiring youths with interest in their studies, as the result of their labour will be proportioned to the earnestness which is inspired by their *desire* to learn. Of all experiences, those which concern the feelings, mental or corporeal, are generally those which most deeply affect the individual, and are therefore those that are most forcibly and lastingly impressed on the mind. Circumstances that induce personal emotions of love, hatred, pleasure, pain, benefit, or injury are rarely forgotten, and they usually respond with remarkable alacrity to the action of memory. Though particular acts may not be specifically remembered, the general idea is in such matters very influential. And this is more particularly the case with infants before they acquire the reason and experience by which feelings are modified in the adult mind.

Persons who have weak memories, and seem able to forget everything else, have generally very vivid recollection of injuries and of other matters of a personal nature ; indeed, the habit of dwelling upon injuries, real or supposed, is a general characteristic of a weak mind. A man who has a very poor stock of ideas may nevertheless feel very deeply, and, having so little else to burden his memory, he seems to cling very tenaciously to these particular possessions.

4. The mental condition at the time of the presentment of a sensation or reflection, of whatever kind, naturally has a great influence on the formulation of the idea of it. A state of *excitement* might either exalt the powers of the mind and serve as a stimulant

to effort, or, on the other hand, it might prostrate or paralyze them. A state of *weariness* produced by protracted exertion is of course a very adverse condition, as its opposite, *freshness*, is eminently favourable to strong perception and effective formulation. Anxiety and all allied conditions of mind have a weakening influence in general, but particularly so in relation to matters foreign to the subject of anxiety. Distress, perplexity, and doubt have a very weakening effect on mental processes generally. Hence ideas formulated under those conditions, unless connected with them, are not likely to be well remembered.

Pre-occupation is a very effective hindrance to the formulation of ideas foreign to the matter which engages the mind. It will even go so far as to shut out other sensations entirely, as in the case already alluded to, and which is given in Note 7 of the Appendix. In less degree it is a very common condition of the mind. Circumstances occurring at such times, and requiring attention, are generally formulated weakly or imperfectly, and, if remembered, will probably be as associations of the particular study or contemplation which was interrupted.

Absence of mind is a form of pre-occupation, and usually denotes the temporary dominion of some habitual and obtrusive thought which, owing to listlessness or indulgence, has acquired dominant influence. This state of mind is often so absorbing as to make its subject almost insensible to external influences whilst the possession lasts.

5. The attendant circumstances under which ideas are formulated, as well as the condition of the mind, are often very influential in the way of association. The fact that, when certain ideas have become associated

together in the mind, one will generally suggest the other, is attested by daily experience. The principle has therefore been adopted in one form or another in nearly all mnemonic systems as a means of aiding memory. Practically, therefore, the subject has reference to two kinds of association, accidental and intentional, the latter having been suggested by the success often attending the former.

Accidental coincidences are generally much more effective than any that can be designed, because they must be striking, or they would not be noticed; and the act of association in such cases is forced on the attention and has not to be artificially created. As aids to the recall of such ideas as happened to coincide with them, they are, therefore, very effective. But they cannot be commanded when their aid is most wanted, and the ideas which happen to coincide with them are often not worth remembering. The moving panorama of life does not so distribute its memorable events and striking phenomena as to suit the mental requirements of individuals, but proceeds on its inevitable course regardless of such considerations, nor can individuals adjust their most important ideas to such occasions.

A very ancient application of the principle of association for aiding memory is the very primitive and, we should think, very effective one which is still practised in some villages in this country, where the youths of the parish are periodically collected and taught the boundaries by being "bumped" against all the boundary posts. As this ceremony is performed publicly, and on some special date or occasion, it is in every way impressive.

Many rules are prescribed in mnemonic treatises

for the application of this principle to general use, and they may be occasionally used with advantage. They are not always easy to follow, nor invariably effective when followed. The artificial nature of the connection impairs its effect. Of the ideas that are to be connected, the one which is to serve as a crutch is not always easy to select, and the choice, as well as the process of associating together the two ideas, requires effort and time. Sometimes the connection, when made, is of less help than hindrance.

6. Repetition is a powerful means of strengthening the force and fixity of an idea. This is of course most efficacious when the sensation or matter is itself repeatedly presented to the mind, but is so in a less degree when the original idea is recalled by memory. Its presentment by the memory, especially whilst the original impression is fresh, affords strong confirmation. Thus, great additional force and depth may be given to the first formulation. Hence an idea becomes very firmly and durably fixed by repeated acts of memory, and may thus be made familiar. Indeed, ideas often acquire by this means an obtrusiveness which is inconvenient or even annoying. Ideas in themselves good may thus become a hindrance to mental effort and be a perpetual source of interruption.

It must be observed, however, with regard to the recall by memory of ideas for the purpose of use or confirmation, that any inherent defects or errors of the original will be confirmed along with the parts which were originally true and complete. Ideas that have thus become deeply fixed and familiarized by repeated acts of memory are apt to be regarded with great confidence. Any defects therein will be maintained with as much tenacity as the rest, without distinction or doubt. With regard, therefore, to ideas which have been strengthened by recall, they must be considered as mere reproductions of the first formulation, and not as though they resulted from re-presentments of the subject or sensation itself. They must be distinguished from repetitions such as would confirm only what is true, and correct what is false or defective in the original idea. We remember how sturdily a young friend once maintained the accuracy of her rendering of a phrase of music. It had become so familiar to her ear by repetition, and was in that form so definitely fixed in her mind, that, on appeal to the text, she seemed more disposed to believe it a misprint than to doubt the truth of the oft-repeated rendering. Memory recalls the ideas as they exist in the mind, inclusive of all errors appertaining thereto.

Repetition is the most effective means of giving force and fixity to ideas, and of making them amenable to the action of the memory. Practical use and application are the most effective kinds of repetition. Language is acquired by children, in the first instance, by means of much repetition, and is afterwards confirmed by constant use. Hence the force and readiness with which it is so soon learned. The desire infants naturally feel to understand what is said to them, the aid afforded to them by the gestures and actions which accompany the words first addressed to them, and the continual repetitions of the words and gestures are the simple factors by which the inestimable accomplishment of speech is so early acquired.

Almost all the ideas which readily respond to the

summons of the memory are those which have been fixed in the mind by frequent use and repetition. Even those ideas which have been first formulated by strong effort of attention seem gradually to become less and less amenable to recall unless revived by use or repetition. They may probably continue in possession and be capable of being revived by force of circumstance or by a very strong effort, but, without repetition, the power of recall diminishes gradually until they elude it, and only re-appear (if they ever do so at all) by means of some incidental association.

A useful form of repetition, which is very generally recommended for its simplicity, convenience, and efficacy, is that of conversing with a sympathetic friend on the subjects it is desired to fix in the mind and to facilitate the recall by memory. When two or more persons read the same book, for instance, either together aloud or separately about the same time, subsequent conversation on its contents and ideas is excellent practice, and answers a double purpose. First, it is a form of repetition of the subject whilst it is still fresh in the mind, and thereby strengthens, fixes, and probably corrects original impressions. Secondly, it serves as an exercise of memory—a faculty which resembles all other of the mental faculties in acquiring strength by use. It has also the further advantage of associating the ideas with the persons and circumstances connected with the discussion.

The familiarity of all those ideas which are in daily use, whether connected with domestic life or with the calling of each person, is such that many common applications of them become mechanical and are actuated unconsciously. They do not, therefore, burden the memory. But those professions and employments which exact an extensive knowledge of details, of which some are seldom used, but must be always at command, tax the power of memory to an extent which causes other acquisitions, new and foreign to the profession, difficult to recall. Men so engaged require to resort to artificial aids, such as written memoranda, and they often complain of their memories without due thought of the way they are taxed and the actual service they render.

The mind may be regarded as a bank on which memory draws, and the foregoing remarks explain the nature of the deposits and show what may or may not be drawn. The function of memory is to recall the ideas that exist in the mind in the form they possess at the time. It cannot make deposits there nor alter any of the items, but it can, with more or less effort, command all the deposits therein. This must be kept in view as essential to a right estimate of the power and a wise direction of the effort of this important faculty.

The science of mnemonics prescribes various methods and contrivances for preparing the subject-matter of Thought in such manner that it may be made as amenable as possible to the summons of the memory. It does not, so far as we are aware, profess to improve the memory itself—that is, the power of recalling all the ideas in the mind—but concerns such ideas only as have been subjected to its treatment. If memory proper could be improved, it would affect the recall of *all* the ideas existing in the mind, and they would thenceforth be all made more available. The rules and prescriptions have not this effect, however, but apply to the formulation of ideas in certain ways and under certain conditions, whereby those particular ideas so treated, and no others, may be made more amenable to recall. In substance, they consist of contrivances for fixing the attention and concentrating effort upon such matters as are to be remembered, in order that those matters may be fixed deeply and durably in the mind. Different systems vary in the form of their rules, but they all tend—(1) to concentrate attention upon the formulation of the ideas to be remembered, (2) to improve as much as possible the opportunity and to extend the time for the process of formulating those ideas, (3) to secure favourable conditions of the mind, (4) to connect with the ideas to be remembered some other salient and appropriate ideas to associate with them, and (5) to repeat the efforts made under all these favourable conditions as frequently as possible.

All these excellent rules, it will be observed, refer to future matter, and not to that previously acquired. They have no relation whatever to the power of recall, but only adapt the subject-matter so treated to be more easily recalled. Naturally these rules will succeed eminently wheresoever they are applied faithfully. Mental treasures stored up by such careful processes will be readily seized upon by the memory, because of the prominent character they acquire by such elaboration. But all the rest of the mental treasures will remain as before. The memory will possess no new power over them. Only those ideas which were formed under the treatment, so favourable and energetic, will be affected thereby. Any ideas that had been previously formed under analogous conditions, if any, will of course continue to possess the force corresponding to these conditions.

Mnemonic rules, such as we have described, are excellent and useful in so far as they are practicable, but they are inapplicable to a great proportion of the matter for which memory is most urgently needed. Hence they cause unreasonable disappointment to persons who have erroneous ideas and expectations regarding their use and purpose. Many persons take up mnemonic systems with the belief that they will thereby obtain new and additional command over all the ideas previously acquired, and they are consequently grievously disappointed to find that no such improvement of the general power is gained. The care and contrivance by which the attention was directed to, and concentrated upon, certain matters tell usefully upon those matters, but have no effect whatever upon others that were not so treated.

Moreover, when rules are reduced to specific form, they do not apply equally to persons of different habits of thought and action of mind. Some rules and devices which would possibly assist certain persons might prove a positive hindrance to others. The safest guide for each individual is to acquire a thorough knowledge of the principles on which the action of memory depends, and then to adapt his practice to those principles by methods suitable to his own idiosyncrasy. A useful exercise is to consider carefully the natural history of the ideas that are most easily and effectually recallable, and then, having ascertained the reasons why they are so obedient to memory's summons, let each one apply the like means to other ideas which he desires to have at command.

We would not be understood to affirm or to imply that memory itself, the actual power of recall, may not be improved, but we do not know of any other means than that which applies equally to all the mental faculties—viz., use and exercise. The exercise of the memory is necessary for every mental process, and in one sense is therefore constant, but the exercise to which we refer, as specially useful, is a systematic exercise. All those ideas that are useful and salutary should be kept bright, and not be allowed to get rusty by neglect. Some of them probably cost a great effort, and may be effectually retained by very little comparative trouble if they are brought out and revived regularly. On the other hand, by neglect they may cease to be amenable, and therefore to be of any practical use. It is pitiable to find how little of all that was so laboriously acquired for certain examinations remains available after a youth has been two or three years in the world. Sometimes this is for want of thought; more frequently it is for want of desire. We know a youth who, on the day he finally left school, dug a deep hole in the garden, and buried all his school-books in the bottom of it! Ideas can only be retained in memory for ready use by being systematically kept in mind by review, repetition, or use. These, therefore, are the means to be employed for preserving all present possessions in a state of readiness for use, or rather such of them as are worthy of being retained. Those which special pursuits or necessities of daily use keep constantly before the mind will be kept by that means; all others of sufficient value must be kept in frequent systematic review, or they will fade from ready recall, and be wholly lost for all practical purposes.

With regard to obtrusive thoughts which may have been allowed to occupy a front place in the mind, they should be promptly and resolutely dismissed, along with all such thoughts as are harmful or useless. Thus may the sanctuary of the heart be kept clean and pure, devoted to salutary and serviceable Thought—fit company for Him who has graciously promised to dwell there, if earnestly invited. Out of the heart proceed the issues of life, wherefore it is to be kept diligently; otherwise, it may be occupied by unwholesome Thought, to the exclusion of other and better guests, or become the playground of wandering, purposeless Thought, suggested by whatsoever passing circumstance or unbidden recollections may take possession of its unguarded gates. Appetite and passion are generally the deputy-keepers of the castle when the true master is off duty.

Considering that Thought is a constant and inseparable companion—to make or to mar the happiness and usefulness of life, to improve or to debase the mind, to cherish the good or to cultivate the evil of the heart and to suggest the conduct—memory, the agency by which Thought is supplied for these purposes, merits the utmost attention that can be devoted to its efficiency.

The foregoing principles are laid down for the guidance of tutors in the training of the infant mind; not that such principles could be either made intelligible to young children or help them to cultivate their own memories, but that they suggest to tutors the essential rules on which that faculty depends, and thus serve to guide them in the work of education. The initiation of right principles of all kinds is a necessary preliminary to their regular cultivation. There must first be a sowing in mental as in agricultural work, and the sowing-time is in infancy. If good habits and right principles be not then initiated by careful training, others will assuredly initiate themselves. The choice is not between doing or not doing. Something will certainly be done; and the choice rests between beginning the work aright or in leaving it, as it is now nearly always left, to chance. The Divine faculty must, we repeat, be either directed and cultivated from the first, or it will follow the lead of chance, which practically means the suggestions of the lower nature. Hence, high intellectual powers, instead of being of common occurrence, are possessed by a few only, to whom the haphazard of the lottery awards a prize. Shining lights—such as Newton, Mozart, Shakespeare—appear like meteors on the mental firmament, whereas they should be as the fixed stars for multitude.

CHAPTER XIV.

ON JUDGMENT.

UNDER this general denomination we include opinions, decisions, and all those operations of the thinking faculty which are required in daily life for the regulation of conduct and for the settlement of principles. The minor matters of constant recurrence, as well as those more important affairs which require deep consideration and concentrated effort, involve the same general principles, and differ mainly in degree and complexity. The nature and value of all these operations depend upon the resources of evidence brought to bear upon them, and upon the manner of the application of such evidence.

The factors involved in all such processes are— (1) the power of the faculty of judging, (2) the stock of evidential knowledge, and (3) the application of the evidence at command to the required purpose, including the selection of such experience, and reference to such standards as the mental resources supply.

Without the power of inference and judgment the senses would be of comparatively little use, for they would then be available only for the perception of objects immediately present and of their present aspects. Memory itself would be of little avail, for its function is to supply the material for the reflective powers. Hence, even the lower animals possess a certain measure of these powers, such as is necessary for their use and guidance, and they exhibit them in nearly all their actions. When, for example, a cat looks behind a looking-glass for one which is supposed to be there, the inference is that the glass is transparent and that there is another cat behind it. When a dog sees his master take his hat, he promptly infers that he is going out. In these simple actions several factors are involved, as previous experience, memory, and inference, the measure of this last being in all cases, both in man and in the lower animals, determined by the limits of the other two. Hence, in young children the power of judgment is developed in proportion to the stock of knowledge acquired and to the strength of the memory. The evidence afforded by watching the exercise of judgment in the lower animals is both interesting and instructive, because it is exhibited by them in a simpler form than by the human subject. The knowledge possessed by the lower animals is confined to their individual experience, whereas men acquire, by means of language and intercourse with each other, far larger resources of knowledge, and their mental exercises are therefore much more complex.

1. The faculty of judgment is necessary in regard to everything that is not self-evident. For example, the existence of a pain is, so far as the mere fact is concerned, self-evident, but judgment is necessary to determine its cause. Of the fact, the sensation is clear and complete, and admits of no doubt, but in order to ascertain the cause, or in some cases even its exact locality, some knowledge is necessary. The power of inference will depend upon the sufficiency

and availableness of that knowledge. The judgment will depend on past experience and knowledge of physiology, both of which may be imperfect or incorrect, in which case the judgment will be correspondingly erroneous. Such, however, as may be the degree of knowledge and the power of memory in bringing it to bear, such will be the resources at command for forming a judgment, and such will be the conclusion. In other words, the faculty of judging depends upon the resources of knowledge and memory in possession. If these be inadequate to resolve the matter in question, the conclusion will be accordingly defective or erroneous. The faculty of judging can do no more than to collate and weigh the evidence afforded by previous experience in its relation to present sensation or reflection.

Some writers believe that the faculty of judging is such that all men would form the same conclusions from exactly similar data. This doctrine may be true, but it is not susceptible of proof, seeing that no two men could possibly possess the same data, nor view them in exactly the same aspect. It is certain, however, that, so far as may be known, the existing differences of judgment and opinion amongst men may almost always be traced to the nature of the data on which they are formed, and to the various relations of individuals to the data. The faculty seems, in fact, to be analogous to a process of weighing, wherein the balance is determined by the weights in either scale according as these are justly estimated and are in their proper places. Manifestly, if the weights, or any of them, be false, or be erroneously estimated, the result will be falsified proportionably. In this connection the humiliating fact must be acknowledged that in every human judgment and opinion not absolutely demonstrable, and in all the principles and actions dependent thereon, there exist inevitable mixtures of error and defect.

It follows that, whatever may be the natural power of the individual mind, the practical value and truth of the judgments formed thereby will depend on the data and their application. The former are supplied and limited by the resources of knowledge, and the latter depend upon habit of Thought, the development of the power of judging, and the personal bias by which the application of the data to the purpose may be influenced.

Habit of Thought affects the manner of exercising the faculty of judging. There is, for instance, a general dissimilarity in the way the faculty is used by men as compared with women, owing to the uses respectively made by them of reasoning and discernment. In regard to ordinary matters, wherein the acute discernment of women is specially applicable, it cannot be said that the habitual reasoning of men gives them any material superiority. The result is generally as reliable and accurate in the one sex as in the other. The difference of habit is rather in favour of the quicker and more ready action of the female mind. In abstract Thought, and in processes of reasoning such as are required in the exact sciences, the masculine habit has a very decided advantage, and judgments thereupon are more sound and exact.

Other circumstances affecting the truth and soundness of judgments are discussed in the sequel. Here we consider the faculty itself.

2. The fund of knowledge which supplies each

individual with the evidence he can command for the purposes of inference and reasoning is composed of ideas of very diverse nature and value. It contains ideas that are absolutely and demonstrably true, along with others in which error exists in every conceivable proportion from the slightly erroneous to the utterly false. A large proportion of the ideas composing the individual treasury of Thought are homemade, the result of reflective and reasoning processes based upon past sensations and experiences. These, of course, all share in whatsoever defects and errors existed in the original formulation of those previous ideas. It is the function of the reflective faculties to extend the range of knowledge by means of reasoning beyond the limits of actual sensation. This is done by inference from data existing in the mind, whereby other ideas are constructed and theories are formed, which vary in every degree from demonstrated truth to the doubtful balancing of speculative probabilities. In other words, the fund of evidence by which each individual is necessarily bound to form all those inferences that depend on memory and reflection is composed of ideas in which the errors and imperfections of original sensations are mingled with those reflections which were based thereon. Such is the edifice his thinking faculty has built up of accumulated sensations from without and suggestions from within.

Hence the power each individual possesses of forming opinions and judgments depends primarily upon (1) the extent, (2) the value, and (3) the nature or character of his stock of evidential knowledge.

(1) The *extent* of the mental stock varies between very widely divergent extremes according to (a) the

age of the individual, (*b*) his opportunities, (*c*) his environment, (*d*) his habit of observing, and (*e*) the power or efficiency of his senses.

(*a*) An infant, commencing life without any ideas whatever, and with faculties as yet undeveloped by use, requires some time for the collecting of even the smallest stock of ideas. Those which are first established concern his appetites and feelings almost exclusively, and these are not fertile in evidential material outside of their own special range. Moreover, as some standard is necessary whereby to judge, this factor is for some time wanting in the infant mind. Hence it follows that the factors or elements necessary for inference and judgment require time for their elaboration. Children therefore judge very imperfectly and often very erroneously at first, owing to the paucity of their evidential material and experience. They are easily misled by appearances, and are too ignorant of their own powers, and of the real nature and relations of things, to form true estimates and conclusions. Their mistakes and misjudgments afford excellent opportunities to their tutors for informing their minds wherein defect appears, for correcting their errors, and for thus training the incipient faculty on which their future welfare so largely depends. Tutors are too apt, however, to neglect such opportunities as not being worth notice, and they not unfrequently make fun of the mistakes children make.

As life advances and the powers of observation and the senses become developed, as the environment enlarges and ideas accumulate, as experience grows and affords better standards, the power of judging becomes more just and accurate, and has a wider range. It follows that, other things being equal, the conclusions and judgments formed in the later periods of life are more mature and more sound than those of youth in proportion to the experienced use of the faculty itself, to the larger fund of evidential knowledge acquired, and to the more appropriate standard by which judgments are formed. It is to be remarked, however, in this connection, that young persons are generally unconscious of their deficiency in the requisite elements for sound judgment, and are consequently even more confident therein than their elders. Indeed, this is to some extent natural, for it is much easier to form a conclusion on a few than on a great many data, and conclusions so drawn seem proportionally clear as compared with others into which many and conflicting elements enter. Much allowance is therefore to be made for the very positive conclusions so common to young persons.

(*b*) Opportunity is a very potent element in the acquisition of evidential knowledge, and is enjoyed very differently by persons in the several grades of civilized society. The general fund differs in such cases rather in kind than in degree. Individuals mix more with others of their own social positions than with persons of different grade. Their ideas and information, therefore, have special characteristics accordingly.

All men, being animated by the same appetites and feelings, are amenable to the same natural impulses; their mutual relations and intercourse are, however, much the same practically in all grades. For those natural feelings and impulses suggest the most influential of the ideas of external experience, and are nearly alike in all men. Love and hatred, friendship and enmity, sympathy and jealousy—all good and evil passions—operate in the same manner in all grades, though they may take different forms in each. The difference of opportunity afforded in the several grades of society is chiefly in regard to such kinds of knowledge as are not acquired by personal intercourse, but depend upon other sources peculiar to each grade.

The sphere of judgment, and its value in any particular matter, differ according to the kind of knowledge possessed by the person who exercises it. Hence differences arising from social position do not affect the power of the faculty, but only the sphere of its operation. They will not affect moral principles, which have the same basis in all men, but only those matters which are peculiar to the social position. Such differences are most evident in early life. In the higher grades, children enjoy great advantages in regard to the acquirement of what is commonly called "learning." In the lower grades, children are necessarily introduced earlier into general intercourse with men and things, and thereby acquire a different kind of experience from that of their more favoured contemporaries of higher grade. As a rule, therefore, the power of judgment is more robust in the self-reliant youth of the lower than in those of higher position, who, mixing less with the world at large, and acting less independently, lack the self-reliance of the former. The training these receive is of a ruder and more practical kind than that of the youth of higher grade, and is better adapted than theirs for judging of the matters of daily life. In these respects the ordinary routine of school training of the youth of the upper classes might with great advantage receive more attention. What is acquired in schools generally is chiefly in the playground, which is in its way deserving of as much attention by masters as the hours of study.

The playground is a little world in itself, and affords fine opportunities for observing character and the working of the mind, especially in regard to judgment. There may be often seen nice calculations of chances and other inferences of a practical nature. The importance of playground proceedings as exhibitions of character is now being increasingly acknowledged, and the active participation of masters in these pastimes gives them influence as well as opportunity, and differs widely from the effect of their presence for mere surveillance. The playground affords quite as good a field for educating the judgment, if used for that purpose, as either the workshop or the office.

Opportunity differs in every individual, and has its influence accordingly of a kind special to each, but our remarks apply to general and broad differences only. Others may be inferred from the known conditions in each case.

(*c*) Environment has a great influence in determining the nature of the fund of ideas accumulated by each individual, and varies in its effects much in the same manner as opportunity. When it is restricted and partial, it tends to produce narrow-mindedness; and, on the other hand, when it is of wide range and varied character, it has the opposite effect, and contributes to expand the mind and to afford a broad basis of observation and experience. Contracted views are generally traceable to restricted intercourse with men and a limited sphere of observation. They are inimical to sound judgment, which requires an ample fund of evidence and experience and effective standards of reference. It follows, therefore, that the most extensive and varied environment which can be commanded is a very important element in education, especially as there is in youths so strong a tendency to over-confidence in their judgments. The reading of pupils should be regulated with this view, and their tastes and pursuits so directed as to keep them from too exclusive adherence to particular grooves, and from the forming of opinions and conclusions on insufficient data. The effect of travelling, as distinguished from a close confinement to a particular locality, tends to enlarge the views by means of extended environment and the acquisition of fresh ideas. These, under the influence of new scenes and circumstances, are particularly impressed on the mind, and are, therefore, generally deep and enduring.

(d) Habit does much towards determining the extent of knowledge. Some persons seem to be all eyes and ears. Nothing escapes their attention except what they intentionally dismiss from their notice. Others make so little use of those organs that they seem to depend mostly on other people's. Again, some persons habitually look closely into things, and are not satisfied with a casual or partial view if a better can be obtained. Their observations are careful and deliberate. Other persons are content with the obvious aspects, and do not look beneath the surface. Their ideas are in consequence superficial and uncertain. Hence the habit of the mind, whether in regard to the manner of observing external objects or in considering internal suggestions, goes far to determine the measure of the knowledge derived from the opportunities and environment. Habit also determines to a great degree the strength or weakness of the ideas in possession; and, as their amenableness to memory depends upon these conditions, their use for the purposes of judgment is likewise affected correspondingly.

Hence the same kind of cultivation which promotes power of memory is equally effective in enlarging the power of judging, as it affords a broader and more certain basis for the exercise of that faculty. The free and forward manner of expressing their opinions which generally characterizes children who have not been repressed affords to tutors excellent opportunities of training their minds to correct methods of judging and of showing them the value of evidence.

(*e*) Amongst the early experiences of infant history many practical lessons are learned, and some of them are very rudely taught on the illusory and unreliable nature of appearances or other evidences of the senses. Who, for instance, has not been allured in his childhood by the glitter of a bit of metal to mistake it for gold? Yet the evidence of the senses, in spite of such warnings, continues through life to inspire unshaken confidence such as they do not always justify. The fact that they may be cultivated, and that they may thus be raised to extraordinary powers of discrimination, is therefore important in its bearing on the power of judging. This is manifest in the practice of such arts as the tasting of tea, wines, and drugs. It is also exhibited in the discriminating power acquired by critics in deciding as to the merits, authorship, and even of the date of works of art. The Hottentot will follow at a fairly rapid pace the spoor or footprints of a man or animal which are utterly undiscernible by the unpractised eye. Many other proofs might, if necessary, be adduced to demonstrate the degrees to which the power of the senses may be exalted by cultivation. But, though the instances quoted have special and limited application, the lesson they teach is worthy of attentive consideration in the training of youth, as it teaches that the use of the senses may be improved by exercise directed to that end. And seeing that so much of the stock of ideas is derived originally from the evidence of the senses, and becomes thereafter an important part of the data on which reflections and inferences have been formed in the mind, it follows that the use of the senses should be cultivated.

(2) Having briefly considered the evidence on which the judgment is exercised in regard to the *extent* of the stock, we proceed to notice the conditions which determine its *value* for the purposes of inference and judgment.

As the knowledge in possession is available only in so far as it is under the control of the memory, it follows that the *value* of the fund of ideas in mind depends (*a*) on the extent to which it can be recalled for use by the memory, and (*b*) on the extent to which it is true.

(*a*) Whatever the actual fund of evidential knowledge may be, so much of it only is of any value for a given purpose as can be commanded for that occasion. Its value as evidence in the particular case depends upon its being in court. Absent witnesses cannot be heard, and do not, therefore, contribute to the decision of the verdict. All, therefore, that is contained in the previous chapters on the subject of memory and control is applicable in this place. As memory supplies the evidence, the power of selecting from the fund the particular ideas appropriate to the case in point is the function of *control*. As this power, when exerted in directing the currents of Thought, determines the personal character in the sight of God, so, when it is exercised in selecting the appropriate evidence by which the practical decisions are formed which regulate the conduct of daily life and intercourse, it determines those issues which concern our fellow-men. These are its highest functions—the most important offices it fulfils in the mental estate.

For the purposes of forming opinions and judgments, the selective power is as important as the power of recalling the ideas required. The selection must be made of just that part of the mental stock that applies to the case, and it must be made *promptly*. It is too late to bring in witnesses when the case is closed and the verdict given. "Ah! I did not think of that," is but too common an exclamation when the discovery comes too late of items of evidence which would have altered the whole state of the case.

It is therefore in its relation to the decisions of the guiding principles of daily life that control exerts its greatest influence. These important issues depend on the efficiency of this faculty, on the completeness of the evidence adduced, its readiness, clearness, and sufficiency. Such power of control is within the reach of all persons who have not weakened their faculties by adverse habits or neglect and indifference. Even such persons may, by resolute effort, do much towards its attainment, but the proper time to begin the exercise and training of the judgment and the faculties on which it depends is in early childhood, when the faculties receive their first bent, and before other habits have obtained influence and created difficulty. The work and the means of carrying it on are then simple. During the period of childhood little more need be done than the correcting habitually and watchfully of mistaken or misdirected efforts which children are continually exhibiting in all their speeches and actions. As the process of learning is necessarily always going on in one way or another, it must either be directed by suitable attention or it will fall under the direction of whatsoever influence may happen to prevail. Thus adverse habits are formed, and the work of the teacher, in endeavouring to undo what has been done, is far greater than would be the simple task of completing a work already well begun.

(*b*) Above all things, it is necessary that evidence should be *true*. False witness in the mental court within is just as mischievous in its way, and perverts judgment as effectually, as in the public tribunals of justice. Hence the value of evidence is mainly dependent on its being correct in fact and true in principle. Seeing, then, that even the senses do not always afford true and complete evidence, and that in all human conclusions there is so great liability to error, care in observing and attention in formulating the ideas which constitute the stock of evidence for all the purposes of life become incalculably important. The facts observed, and the reflections founded on the ideas formed of them, are like entries in a ledger. If true and correct, they yield only true results; but one false entry may vitiate many accounts and affect many calculations. In like manner, erroneous ideas in the mind vitiate the conclusions into which they enter, and may therefore lead to serious consequences.

Indeed, the history of nations and the experience of individuals furnish abundant examples of such consequences arising out of erroneous ideas where no ill intention existed and where no error was suspected. The most terrible calamities that have befallen civilized nations have originated in erroneous doctrines and theories. Bigotry and intolerance, for example, which have consigned to cruel death and martyrdom hundreds of innocent persons for slight differences of opinion, result from mistaken ideas of authority, whereby a fallible dogma or erroneous interpretation is assumed to possess Divine authority. Error is thus propagated and truth violently suppressed. Such forms of error are not confined to any particular branch of knowledge, though religious differences are probably the most virulent. Error finds many ways of affecting the conduct of individuals in unsuspected directions, and spreads its baneful influence through all social institutions. Men, overlooking their individual liability to error, seek to impose their opinions and conclusions upon their fellows, and thus, unintentionally perhaps, to overrule truth, in their zeal to maintain the false views of which they fail to perceive the fallacy in themselves.

In considering the multiform operations of judgment, and the amount of error and imperfection that are inextricably mingled with conclusions based upon opinion, it is curious to remark how much of what is so dogmatically insisted upon is based on fallible dicta which have happened to acquire the form of authority without any real claims thereto. Such dicta, when crystallized into creeds of religious belief or theories of science or philosophy, exercise boundless influence on mankind.

Hence various forms of gross superstition have prevailed for ages and swayed the conduct of many generations of people, who were thereby enslaved pitiably and demoralized cruelly by systematized falsehood. Built on conjecture, in ignorance of the true aspects and nature of things, these formulated beliefs and theories have grown gradually into stupendous edifices of error, and have degraded the better parts of the race of mankind. How much of such mental slavery is still in force in this nineteenth century may not be fully known for a century to come, but, when it becomes exposed, it may not look much less degrading than do the historic fictions and fables of the Middle Ages to the present generation.

The most hopeful feature of modern Thought is that all theories and principles are being submitted to rigorous scrutiny, and that they are being maintained only in so far as they are supported on a stable foundation of truth. Every historic fact, scientific principle, and religious doctrine is now referred to its rightful claims, and to them alone. Flagrant errors having, as history attests, so long maintained their ground and enthralled mankind for ages, ancient usage has ceased to make doctrine venerable, for those which have done their work most cruelly have often persisted most tenaciously.

Baconian methods have superseded dogmatic and chimerical theories in every department of knowledge.

Religion, which by reason of its high pretensions has suffered the most, has also nevertheless been the greatest gainer by the application to its doctrines of the searching method of modern analysis. Hence the mental atmosphere has been cleared of much prevailing error of former times, and judgment is exercised upon more truthful data. The claim of infallibility set up by religious systems which refuse to admit the possibility of error, has proved the most destructive instead of being the most powerful support of authority. Such claim necessarily exposes the whole fabric to ruin by the confutation of any, even the smallest, part. Hence the mass of priceless truth contained in Scripture has been held to rest for its authority on the maintenance of trivial errors of translation or copy, such as were inevitable in all human work. The sacred truth is now no longer subject only to the inimical scrutiny of unbelievers, but research is conducted in conformity with Divine command, by those who are most deeply concerned in ascertaining their truth without regard to preconceived or borrowed beliefs, and with a single eye to the maintenance of that only principle which can ever be durably sustained—the truth.

The foregoing remarks bear directly upon the history and training of the mind in relation to individual experience, for there is no other way in which the far-reaching and powerful influence of error in the data upon which human opinions and judgments are based may be so fully recognized as when they are exemplified in such matters as we have quoted. In such widely extended operations and stupendous results, the vast influence of false principles are most plainly and indisputably manifest. But error is of even greater moment to men individually in its results on their own lives and conduct than it is in those more general forms in which it affects the masses. These concern the character and progress of an age, but those affect the personal experience, the status of the individual and his relation to his Maker. These determine popular beliefs and historic prejudices, but those decide his life and conduct, the sacred trust for which he is personally and individually responsible. Hence, personal opinions and judgments, the determinants of a man's conduct in life, are of far greater import to him than are the current opinions and popular theories of his day, whether they relate to religious doctrine or the theories of science. It is therefore the personal view of the question for which the foregoing illustrations are adduced, in proof of the vast influence of error as an element of Thought. With such evidence in view, the value of Truth in all the elements of Thought needs no other enforcement.

(3) The *nature* of the knowledge in possession regarded as the evidence on which all opinions and judgments must necessarily be founded, is of two principal kinds :—(*a*) particular ideas ; (*b*) general principles in their specific character ; or (*c*) in their relative functions as standards, to which all subordinate ideas are referred. The office these last fulfil is to give the specific value to judgments and opinions. For example, the idea of wealth will depend in each individual on the standard of his

own personal experience. An income of one hundred pounds a year would be wealth to a beggar; but to a millionaire it would seem abject poverty. Hence, in all matters of comparison between different things or different aspects of the same thing, ideas are re-referred to some personal standard resulting from experience.

(*a*) Particular ideas of things, or of their relations and qualities, differ in the degrees of their importance as elements of evidence according to the extent to which they enter as factors into the opinions and judgments of their possessor. Like items in a ledger, they may affect only one or two accounts, or they may be of a more general application, and affect a great many. An imperfect idea of distance, for example, owing to defective vision, inflicts many inconveniences, whereas a still more imperfect idea of a particular plant or place may cause little if any mistake of a practical nature.

As it is not possible to foresee to what uses any idea may hereafter be applied, a habit of care in observation, even in things seemingly unimportant, such as to ensure a reasonable degree of correctness, cannot be too highly commended. For example, a traveller notices a particular practice which obtains in a place he visits. It attracts his attention for the moment, because perhaps of its novelty or of its utility. Years pass on, during which the idea of this practice does not recur, until at length its applicability to some present purpose revives the recollection, and he desires to take advantage of his former experience. Supposing his *habit* to have been one of care and attention, the idea as revived will enable him at once to adopt and apply the practice. But, if otherwise, he may in vain tax his memory, for it can only recall the casual and incomplete idea he formulated, and he will either have to forego the advantage or to puzzle out the deficiencies by laborious trials. Hence, the far-reaching influence of habit, even in matters to which no particular importance seemed to be attached. A careful habit of Thought makes all the difference, in many cases, between a person whose opinions deserve credit, and another on whose carelessly formed ideas no reliance can be placed. With reference to the vital functions of opinion and judgment, all we have so strongly expressed in previous chapters respecting careful habit applies with its utmost force in this connexion.

(*b*) The most influential data on which judgment is exercised are those of *general principles*. These have wide application and corresponding influence. Such are the ideas of duty, moral obligations, and religious opinions, which regulate general conduct as well as particular judgments, and give the tone to the life and character of the individual. The basis on which such general principles rest, and from which they derive their authority and influence, cannot be too jealously scrutinized in all their bearings. Every man should have a reason of *his own* for the hopes he indulges, the love he bestows, and the general principles on which he acts. And the reason should be adequate to the occasion, for whatever his ideas may be, he will never act courageously and consistently in cases of difficulty if he doubt the truth and authority of the principles which demand his obedience ; nor can he be said to hold them safely, whose tenure is that of borrowed creed or theory.

The prolific nature of doubt and error in reproducing their like is in no other way so manifest as in those principles and obligations imposed by formulated doctrines, whether they be political, social, or religious. As a rule these are borrowed, or adopted on trust, without due regard to the foundation on which they rest, and to the momentous fact that the borrower is personally responsible for the influence they practically exert upon his own conduct. It is an easy method of evading the trouble of research to borrow opinions and principles from others, and thus to have one's political, social, and religious principles ready made. But though present labour may be thus saved, future responsibility is not thereby escaped. Ready-made religion has been very effectively denounced by Professor Drummond in his bold essay on "Natural Law in the Spiritual World," and merits all the denunciation with which his book abounds. The moral cowardice and mental weakness implied in entrusting such vital interests as those of moral duty and religious exercise to the keeping and direction of erring mortals, are a sad characteristic of a kind of bondage which is too prevalent.

The useful rules and principles of grammar, arithmetic, history, geography, &c., legitimately claim the attention they usually receive in the process of education, but considering the importance of those moral principles which regulate character and conduct, these are surely entitled to a degree of attention commensurate with these vital interests involved. In the education of youths, however, parents generally regard the religious instruction to be given with reference rather to sectarian distinctions than to the broad general principles of Christianity. We lately heard a youth severely scolded for having bought an article at the shop of a Churchman, he being a Nonconformist! Though such petty bigotry as this is probably a rare extreme, it is the exponent of a feeling of narrowness which is by no means uncommon. Amongst the equipments necessary for a youth who is about to enter upon an independent career, the elements of judgment, mercy, and faith, should not be sacrificed to the tithes of mint, anise, and cummin.

The importance of general principles in reference to the exercise of the faculty of judgment entitles them to primary consideration in education. We do not undervalue the use and convenience of special rules, but these should be referred as corollaries to the broad principles on which they are founded. As ready means of applying those principles, rules have a high value, but the grand comprehensiveness of general principles should not be frittered away into little bundles for special use.

(*c*) High and just standards are essential to adequate ideas of excellence in regard to all judgments.

In regard to *moral* excellence which determines the most vital interests of mankind—namely, the conduct of this life and the aspirations for a higher and spiritual state of being, the Christian possesses a perfect ideal in the simple and beautiful life of the blessed Redeemer. He who imitates that sacred example, as the rule of his life and the basis of his faith, cannot fail to excel in all that pertains to virtue in its largest sense.

Standards of *knowledge* generally are shifted from generation to generation, according to the progress that is made by each. A simple illustration of the practical effect of standard in regard to judgment is seen in the admiration a child will waste on very poor performances, for want of better knowledge and larger experience of attainable merit. This leads us to remark upon the very great improvement that has taken place during the last half-century in the style of childrens' books and toys. The barbarous drawings, daubed with flaring colours, which formerly adorned the books which were to educate the taste of the rising generation, and the hideous toys, models of ugliness and misshape, which parodied the works of Nature, have been superseded by works of art. Hence, children of this period have far higher and more just standards of taste in all such matters established in their minds from their infancy, than those which were enjoyed formerly by any but the upper classes of society. And the general public enjoy a corresponding advantage by means of museums and public exhibitions which have been made accessible to all classes. Hence, those who half a century ago had no knowledge of any but barbaric caricatures of art, may now procure cheap copies of the best works of the greatest masters. Thus have the standards of art been raised in the mass of the people, and public taste has been thus elevated far above the former level, of which the recollection survives only in specimens of a past generation.

The same cannot be said of the cheap literature of the day, in regard to which the standards of taste and moral tendency have not been cultivated. The rage for sensational works of fiction, which has grown with indulgence, seems, however to be working itself out. As it becomes higher in flavour it gets nearer to that stage of corruption when it will become intolerable. Surfeited appetite will avenge itself, and will crave more wholesome food in disgust of the trash formerly evoked. On the other hand, the standard of sterling literature has not been so debased nor its supply reduced, though it makes less figure amongst the preponderant bulk of trash that issues from the press.

Fashions and particular characteristics of each age are due chiefly to the varying standards which from time to time gain influence with the public mind. Scientific, intellectual, and social standards would each and all supply examples to attest the influence these factors exert on the opinions both of individuals and people of different periods, but for our purpose it is not necessary to refer to them. The foregoing examples suffice to show how influential are the standards employed in determining the character and value of the judgments formed thereupon. Hence, in the training of the thinking faculty in youth the subject deserves particular attention, with a view to their being instructed in the true principles and practice of estimating data and of forming correct opinions and judgments generally.

3. Having now considered some of the points which refer to the *extent* and *value* of the stock of ideas which constitute the evidence on which judgments and opinions are formed, we proceed to consider the *application* of such evidence to its appropriate use.

A mass of evidence, however pertinent and true, and however obedient to the summons of memory, does not of itself compose a judgment. That is a separate act. Evidence needs to be analysed, classified, and applied to its purpose. The items require to be arranged and estimated. And this must be, and actually is, done either by haphazard or on principle whenever a verdict is obtained. The nature and value of the conclusion will in every case correspond with the manner in which this work has been done. Every opinion and judgment is formed upon data existing in the mind. It neither comes of itself nor by any foreign agency, but is the result of a process performed in *some* way by the mind, and most people hardly know how. They have been used to "making up their minds" all their lives, without reference to any known principle or method, and are consequently sensible of much perplexity whenever an unfamiliar matter of importance has to be decided. They do not address themselves to the work on any orderly plan, but trust that the right evidence will come to their help, and they thus arrive at a decision in which they have little confidence. Probably after-thoughts soon come to disturb the first decision and produce vacillation and embarrassment. In the end, after much desultory mental debate, the result is full of doubt and uncertainty. The process, when so conducted, resembles what happens when men attempt to do other things without knowing how to do them properly. They have the materials, but do not know how to use them.

Give a man who does not understand shoemaking the materials of which to make himself a pair of shoes, and he will not know how to begin. After one or two false starts, spoiling a great deal of material and wasting much time, he will end with a wretched bungle. Compare this proceeding and its result with the work of an adept, who, in half the time, will have completed one or more pairs easily and creditably. These processes represent the difference between mental operations done ignorantly and unsystematically as compared with others done deftly and on system. But men do not *try* to make shoes, to build houses, or to plough fields, without previous training. Their bungling work would be *seen*, and would expose them to ridicule. Their bungled opinions and vacillating judgments, on the other hand, are not *seen*. They are concocted in the secrecy of their own breasts, where there is no one to ridicule the process or to criticize the result. Hence it seems to be considered quite immaterial how such mental processes are effected.

Seeing that no one would think of attempting any of the ordinary avocations of life without some form of training to qualify him to do the work correctly, it seems strangely inconsistent that mental work should receive no such training. The inference is that it is either unimportant or impracticable. The former alternative needs no argument, seeing that decisions and opinions of all kinds are of daily, almost hourly, recurrence through life, and that many of them affect rules of conduct as well as material interests. Adequate training is at least as necessary for actions involving such issues as for the ordinary avocations of life. The equipment of a youth for his independent career should not be less complete for the processes of Thought, opinion, and judgment, than for bootmaking, carpentry, and the like. Yet, practically, no systematic preparation is made for this work, which is therefore bungled over by clumsy processes, such that if they were not concealed from view they would excite contempt and ridicule. Being kept secret in the privacy of the individual breast, however, no instruction is deemed necessary, and the fact remains true as it was stated by Locke, " that the last resort a man has recourse to in the conduct of himself is his understanding."

The training of the thinking faculty if it were recognized as an essential part of education would be found to present no difficulties more insuperable than other kinds of training, to which the youth of both sexes already submit for other objects of far less importance than that of right thinking. Great attainments reward efforts systematically directed in other matters, and great benefits accrue both to individuals and to the community therefrom. Most of these are begun in early life, but the comparatively few attempts that have as yet been made to educate the thinking powers have been addressed to and intended for the adult mind. Hence, they bring their very hard lessons to bear upon minds which have already been moulded into settled habits by the chance processes of untrained effort. Their minds, already so educated, are not generally amenable to a new process of education.

Prejudice is the natural fruit of the untrained intellect, which judges imperfectly and erroneously for want of correct views of the value of evidence and of its application. The stupendous edifices of superstition, error, and prejudice, such as the astrology, alchemy, and witchcraft of former ages, which all crumbled into the dust when Bacon convinced men of the only true and stable data on which judgments could be safely formed and truth established, prove conclusively how efficacious are right methods and principles in regard to mental action. Yet the essence of Bacon's method can be made easily intelligible to the minds of very young persons, and may be applied as general principles in the education of the thinking faculty.

The considerations adduced in the foregoing part of this chapter show that in the education of children there are two principal objects to be accomplished, one being to *inform*, the other to *discipline*, the juvenile mind; one serves to supply a suitable and sufficient stock of knowledge, the other to train the mental powers in the right methods of using or applying such knowledge. The necessity of the former is universally admitted, and is probably as well and successfully done in the usual routine of education as is compatible with the almost total neglect of the latter. Yet the process of informing the mind and storing it with the knowledge necessary to be used in the purposes of life, affords the very best means and opportunities for teaching such use, and the processes appropriate thereto. The best stimulus to learning is a desire to know. This is a chief reason for the facility with which language is learned. The natural desire to know what it all means is the impulse which so effectively promotes the acquisition of language. If learning be a manifest means to a desired end, there is a strong inducement to learn; and in *so* learning, with a definite end in view, there are implied both learning and discipline. The relation that subsists between an idea and the uses to be made of it is the most practical and elementary form of introducing and understanding the processes of applying it to useful purpose. On the other hand, the study of subjects that have neither present interest nor future promise cannot be made palatable nor inspire desire. They are at the best dry, and their pursuit drudgery. Hence it is that so large a proportion of what is best known and most available for use is that which was acquired under the simple stimulus of a desire to know it.

Before the age when reason can be invoked, that is, before a fair use of language has been acquired, children learn by means of imitation, and can then be taught only through the medium of example and correction. Other forms of teaching are never resorted to by wise tutors before speech is fairly acquired. All that is learnt previously is simply imbibed of their own choice. Unfortunately, as they exert no selective power they imitate as readily things not intended to be imitated as others that were proper. Afterwards, when teaching is commenced, desire to learn is so essential to success that it is quite as important to excite that feeling as to adjust the form of the lesson.

Euclid is generally regarded as a very useful means of instructing the reasoning faculty. It no doubt sets forth in practical form some of the principles and processes of forming correct conclusions, but apart from this particular use it does not inspire interest in the minds of most youths, who consequently drag through its seemingly purposeless propositions wearily and unprofitably. But to a youth of fairly advanced acquirements, who has a definite idea of the purpose for which the study is commended to his consideration, it becomes invested with a special interest, and he will in nine cases out of ten get over the asses' bridge without ever knowing of its existence. Still, though eminently useful, Euclid is not a sufficient nor altogether a good means of cultivating the judgment. Referring as it does to positive demonstration only, it is not applicable to the great mass of human affairs, which do not admit of absolute proof. These require rather a power of estimating degrees of probability and the value of evidence than the close reasoning through a chain of certainties. It is rather liable, therefore, to create a doubting tendency in regard to such matters as do not seem to have such rigid certainty. Moreover, Euclid comes too late to *initiate* the training of the judgment, which should be begun by the corrective process. It serves better to confirm and elucidate previous ideas of the process than as a first step to their acquirement. There is besides a specialty about Euclid which prevents a ready apprehension on the part of the student of the general principles of reasoning to which it owes its chief educational value. There is a tendency rather to exalt the science of mensuration than to teach the art of just inference.

A curious and instructive illustration of the estimation of evidence is given by Babbage in his ninth Bridgwater Treatise. A problem submitted for solution was to determine, by means of the indications of a certain machine, what was the general law of its movements. The machine being set into motion, the several observers who were to solve the problem noted the records of the index. After one hundred repetitions of a progression by a unit, one observer was satified that it would so continue, and he retired. Others not so easily convinced proceeded to note the further indications. At the completion of the one thousandth term, still in the same order, all but one regarded the law as established. He only persevered, and discovered that at 100,010,002 a new order of indications occurred. Yet, from the first, the change which then took place was a necessary consequence, or a law, of the mechanism. This illustration serves to show the different estimates men form of the value of evidence bearing on a point to be ascertained. One jumps to conclusions prematurely and impetuously, others with more patience, but few with sufficient care to arrive at any but the most obvious decisions. The care and patience exercised by the last-mentioned is not to be confounded, however, with that unhappy condition of mind, in which doubt and indecision dethrone judgment in all matters that do not admit of absolute demonstration. As such matters form a very small and comparatively unimportant part of those which constitute the experiences of mankind, persons so afflicted have but a small basis on which to exercise the power of judgment.

Judgments and opinions are subject to considerable influence from other considerations besides those of evidence. Such, for example, are the influences of self-interest, temper, prejudice, state of health, &c. These all tend to disturb the equilibrium on which the truth of the result depends. The seductive force or repellent influence of such influences is seldom appreciated or even known to those in whom they operate. They are false weights unconsciously put into the scale, and vitiate the balance. *Self-interest* perverts the mental vision which reads the weights untruly or adjusts them falsely. The wish is often father to the Thought. *Temper* is also a powerful agency in falsifying the weights and corrupting evidence. A calm and undisturbed state of mind is a necessary qualification for sound judgment on whatsoever subject it may be exercised. The state of *health* also reacts on the mind and gives a false colouring to the elements or a partial view of the evidence. The exaltation due to stimulants is said to sharpen the powers of mind, as in Pitt, Sheridan, and others, who exhibited their highest powers under that influence. Still, we think that he who appealed from ·Philip drunk to Philip sober showed a sound knowledge of the working of the human mind.

Prejudice is probably the most general and influential of all the forces that tend to pervert judgment. The existence in the mind of previous conclusions, based on borrowed ideas or on insufficient evidence, or perverted views of evidence, interfere to an incalculable extent with the due exercise of judgment. The influence of prejudice is most widely exerted in matters of general principles, such as have already been alluded to as affecting the ideas of people generally. Astrology, some forms of superstition, and witchcraft have for ages inflicted error, and prompted to crime millions of men, and still linger in remote parts of the earth to exercise their maleficent powers. Still, it is rather with individual prejudices than with those of an age or nation that we are here concerned, and the subject is too wide for discussion in these pages. We conclude our remarks with one illustration derived from our own experience. Conversing with a most worthy and highly gifted prelate, he maintained that the fossil remains of animals embedded in the rocks must have been created there. This excellent man preferred to adhere to the uncertain interpretations of bible chronology rather than to disturb his preconceived ideas by accepting the plain certainties of science. Here is a fair example of thorough-going prejudice, operating powerfully in a highly cultivated mind. Such prejudices must continue to operate more or less generally until the practice is fully established in religious and secular matters, as it already is in science, of giving all verdicts according to reliable evidence.

A prolific source of error in judgments of all kinds is the imperfection inherent in language, the medium of interchange of ideas. The careless use of terms, to which a good deal of ambiguity has been attached, by either their abuse or by the deficiency of language to supply other terms more precise and definite, gives rise to dispute and difficulty in almost every argument which requires precise formulation. This subject is so fully treated in many works of authority and excellence that it needs no more than mere mention in these pages. Of course whensoever the necessary imperfections of language are further aggravated by the inevitable difficulties of translation, as in the Holy Scriptures, the liability to erroneous inference is much increased, especially as there do not exist in all languages exact equivalents of the terms of each. Hence the introduction of terms from one to another is of constant occurrence. Numbers of words in our own language have been incorporated therein from others, because no equivalents were in existence.

It would be manifestly vain and foolish to attempt to teach the principles embodied in this chapter to a child in the form of verbal instruction. Nevertheless, as they are the essentials of sound inference they must be acquired at some time, and in some manner, if the all-important function is ever to be rightly and effectively exercised. And it stands to reason that the sooner they are learned the better will it be for the learner, for it has already been shown in our chapters on the natural history that, rightly or wrongly, children do exercise their faculties in inferences, opinions, and judgments, and it is equally certain that their decisions, whether right or wrong, actuate their conduct in many ways. Hence in their first conclusions they afford to tutors the best opportunities for judicious correction, which, given in accordance with the foregoing principles, will practically teach them by their application which is the best possible way in which any principles can be taught. In our chapter on training the infant intellect, simple but effective rules are laid down by which the principles of right-thinking may be inculcated and right habits of Thought initiated. In the more advanced stages of youth the faculty of inference and judgment might be cultivated by means of exercises involving simple, practical propositions, the written answers to which would test the student's powers of reasoning, and would afford his teachers the means of instructing him in the application of true principles.

CHAPTER XV.

ON INHERITED CAPABILITIES IN THEIR RELATION TO THE ACTUATING PRINCIPLES OF CONDUCT.

THE question as to what mental characteristics, if any, men may inherit from their parents, is one of paramount interest in the study of mental operations, for it determines whether and to what extent, character and attainments are attributable to individual experience and effort, and also whether, and to what extent, men may derive their knowledge and other mental characteristics by descent from parents or ancestors.

Indirectly the question has important bearing on the degree of moral responsibility properly appertaining to individuals themselves, and also to that which devolves upon tutors. It has moreover a special interest in its relation to the exercise of individual free-agency.

There is no dispute respecting the fact that the physical organism of a babe is dependent on inherited properties and characteristics for the condition in which it is born into the world and becomes an independent living being. Nor is there any doubt that the brain, which forms a part of that organism, is similarly dependent, and that it is the part by which mental action is connected with that of the other organs. In other words, the brain is the acknowledged seat and centre of the mental functions of the physical system, and as a part of the organism it is derived from parents.

It is further admitted, that the physical and mental powers in men differ in degree between very wide extremes, both as regards power and sensitiveness, in different individuals. Whatever differences there may be in other respects, they are in all cases alike in this one respect—that they are dormant until their rightful possessor is in a condition to exercise them. There is a manifest absurdity in the idea that any of the organs had been used or had fulfilled any of their appropriate functions before birth. No one ostensibly supposes that any of the organs are used until the babe himself acquires the power which life imparts to use them for himself. Yet the doctrine of innate ideas, on which inherited temper, for instance, inevitably depends, does actually involve the conclusion that the brain had already begun to fulfil its functions before it could possibly have possessed any power of either sensation or action! The doctrine of innate ideas has long been discredited in theory, but nevertheless, in practice, it survives in the popular belief that temper and other mental feelings are inherited, and such belief implies that pre-existence of the feelings which constitute temper.

The nature of ideas, and the manner in which they are acquired, preclude the possibility of their being inherited, because they are the records and representatives, in each individual's mind, of his own past sensations and experience, and are produced thereby. The manner of the formation of ideas is the same in individuals of every capacity.

Whatever may be the degree of his natural power, every one knows by his own experience that he acquires his ideas by the use of his senses and by reflecting on past sensations and ideas. He knows therefore that ideas are so produced, and that they remain as records or images after the objects or reflections have passed out of notice. Such being the nature and origin of ideas, it is manifest that they could not have been formed in the mind previous to the epoch when life first conferred the power of feeling, thinking, and acting. It is also evident, from their nature and use, that if it were possible to infuse an idea into the mind otherwise than by the natural and only known method of personal experience, such an idea would have no significance. Not being a record or representative of a personal sensation or reflection, it would be unmeaning and unintelligible, because it would have no connection with any known thing. It would be like a word of an unknown tongue, without meaning or utility. The existence previous to birth of any such feeling as temper, which is in its nature a relative feeling, is therefore manifestly impossible. The doctrine of innate ideas is a futile device, intended to meet difficulties which it fails to explain, and which may be sufficiently accounted for by other causes of familiar nature.

Hence experience proves that the human organism, as inherited at birth, possesses certain capabilities, which vary considerably in different individuals; and that the power of using the limbs and organs, including the brain, cannot be exercised until the babe receives the breath of life which confers that power. The organism, with its measure of capabilities, is inherited, but the idea that it has either acted or experienced in any way anterior to its own independent life, is opposed to both reason and experience. The babe has keen susceptibilities at his birth, but could not have any ready-made feelings or experiences to begin life with, because he had till then no power of sensation. He also possesses mental faculties, but, as these can only be exercised by himself, they are unavailing until life confer the power of using them. He cannot therefore possess ideas ready formed. Action and sensation are functions of life, and cannot therefore exist antecedently to life.

Our contention is, that body and mind are equally naked when life animates and sets them into motion, and that till then the only equipment for independent life consists of the organs and their respective capabilities.

Seeing that the usual method by which ideas are acquired provides so well for all the mental requirements, that it is so well known, and that it is the only means of which men have any actual experience, we have to consider why any other means should be sought. The reasons appear to be:

(1) That certain manifestations of temper, and other relative feelings, often occur in very early life, for which no adequate cause is apparent.

(2) That certain persons distinguish themselves by extraordinary attainments, which are remarkable, either by their development at a very early age, or in a very high degree, such as to suggest the possession of more than the regularly acquired powers.

As no adequate causes for such characteristics are apparent, resort is had to the alternative that they were due to inherited gifts. This theory owes much of its vitality to the very convenient excuses it affords to tutors for what would otherwise be referred to their neglect or example. To the victims of ill-humour it serves to exculpate themselves and to lay the blame on their ancestors. To persons of indolent and wayward habits, it gives a plea of deficient talents to excuse want of application or of right direction to their efforts. A theory which serves such purposes to such persons will not be relinquished readily, and we hardly hope to prevail with them.

The only positive evidence adduced in support of the doctrine in question, is that of phrenology, which, however, in the present state of the science, is not very reliable, and were its data more accurate, the evidence would not be conclusive. If the particular function of each part of the brain were clearly ascertained, and if men necessarily exerted their powers in proportion to the avoirdupois weight, cubical contents, or special sensitiveness of those parts of the brain, the evidence of phrenology would have much value. But such is not the case. Men do not generally so employ the powers they obviously possess. As to the *special* capabilities we are now discussing men are often mistaken respecting those they imagine they possess. It is indeed common for them to be unconscious of the powers they actually exhibit, and to plume themselves upon others of which their neighbours perceive they are deficient.

We now proceed to consider generally the theory

of inherited mental characteristics, such as temper and other states of mind, extraordinary special capabilities, &c., and will endeavour to demonstrate:

1. That life itself is not inherited, but only an organism capable of receiving life, and therefore as life is not derived from parents, but follows birth, neither the mind itself nor any temper, state or characteristics of mind could exist before birth.

2. That men do not necessarily employ the powers they actually possess, nor is the use of their powers generally regulated by the relative proportions of their capabilities. That they are often unconscious of the possession of particular capabilities, which are nevertheless evident to their fellows, and are otherwise mistaken in their own estimates of them.

3. That extraordinary mental attainments do not appear to be generally referable to special capabilities as distinguished from general power.

4. That the theory of inherited *special* capabilities, as an actuating principle of conduct, is not borne out by the facts of daily experience.

5. That as man is endowed with natural faculties wherewith to acquire knowledge and to act for himself, it is not reasonable nor consonant with analogy that his natural powers should be superseded or their functions anticipated by other means.

1. Our first proposition depends mainly upon the way in which life originates and the organism becomes animated. This therefore requires first notice, as life is the first essential for either feeling, thinking or acting.

As an electro-magnet has no magnetic property inherent in itself, but only a susceptibility to receive it under certain conditions, so in like manner the germ of a plant or animal does not itself possess inherent life, but only the property or susceptibility of receiving it by means of suitable conditions external to itself, such as in the order of Nature have been ordained. St. Paul says that the seed is not quickened except it *die*, whereby he expresses no metaphor, but states a scientific fact—viz., that until animated by the quickening process the seed is dead.

Life is essentially an active, energetic, natural force, which, in the sphere of its operation, acts powerfully and overcomes other forces, which in its absence exert their powers in very different directions. When the vital force ceases to operate in any living organism, the component elements of such body speedily fall under the influence of those other forces, and decomposition ensues more or less quickly. So long as the vital force operates, the living organism is maintained by means of a cycle of processes which renew its tissues and sustain its powers. But when these active processes cease, the component elements of the organism become subject to other forces, and the organism is decomposed. Thus living is a state of activity, and life in its nature an operative force. Wherefore latent or inert life involves a contradiction.

Is it well known that seeds that had been buried in tombs in Egypt have retained their susceptibility to the quickening influence of the conditions on which vitality depends, after thousands of years of complete inactivity. In regard to these seeds, many of which have actually germinated, it is certain that their germs could not have been kept in a truly living state without consuming the substance of the seeds in which they were contained in order to sustain a vital condition. On the other hand, the supposition that they maintained vitality in a latent form would be a contradiction of terms. The seeds and the germs they contained were, as St. Paul says, dead, but capable of being quickened.

The capability of receiving life is not one whit less marvellous or inscrutable than life itself, nor is Almighty power less manifest in ordaining the process of quickening than it would be in imparting life by a direct interposition. So far as is yet known, the only organisms that possess the capability of being quickened derive that property by the agency of parentage. Still, the property or susceptibility of receiving life is not inherent life, nor does the parentage confer life, but only a fitness to receive it. There is a clear distinction between the vital force which animates and the organism which is thereby animated. There is in fact a duality involved in living organisms of which the two elements are different both in their nature and origin. The germ is the offspring of parentage, but the quickening power or vital force has another origin, being induced by appropriate conditions, ordained by the same Almighty power that ordered the production of the germ and its envelop by the action of parentage.

In plants the germ is encased in a seed which, when completely developed, is separated from the living organism in which it was formed, and remains inert until the conditions necessary for quickening the germ supervene and endow it with the life for which it is specially adapted. Those same conditions do not avail for any other organisms than those so naturally qualified. What the conditions are is not certainly known, but those which are apparent are air, moisture, heat, and light. All these, in some proportions, are necessary, but whether there may not be others remains to be ascertained. Until those conditions supervene, be they what they may, the seed remains, like the magnet without its electric current, inert.

In the animal economy the processes are analogous to those of the vegetable kingdom, but differ in regard to the exercise of the maternal function according to the manner of birth proper to each class of animal. The fœtus before birth shares the vitality of the mother, and forms part of her substance, as does also the seed of the plant until the flower or the plant itself dies. The child has no life of his own until after birth. There is not unfrequently some difficulty and delay in the inception of life after birth, in the case of the babe. The involuntary functions of the heart and lungs are in some cases dilatory, and require assistance. The quickening agency in the case of the animal offspring seems to be air, but though this alone may be apparent, there are so many subtle agencies in the world that it would be bold to assert that no other is concerned.

Seeing then that life is not inherited, it follows that thought and feeling, which are functions of life, could not be inherited, nor could any state of mind exist until the mind itself is developed by vital action. Even those persons who hold the possibility of life being latent, will perceive that if *latent* it could not have operated in the production of Thought or feeling.

The question as to the nature of what is actually inherited is thus narrowed into a comparatively small compass, for it is evident that at the time of birth the babe possesses an organism, with an organ or appliance for thinking, but that he could not possibly have any ready-made Thought. That would be as absurd as to suppose a violin with ready-made tunes in it. Hence it is shown that the only properties men can possibly inherit are the means or apparatus necessary for living, thinking, and acting. No such thing as temper, or other state of mind, is possible prior to life. They inherit an instrument for thinking, but no Thought already done by it, nor any ready-made Thought existing in it.

It follows from the foregoing arguments that all the ideas, feelings, and states of mind of which an individual is conscious, must necessarily have been acquired by himself, seeing that they could not possibly have existed before birth. And it has also been shown that if such ideas were possible they would be unintelligible, unmeaning, and inoperative, because they would not represent any known experience. Having thus considered the theory of inherited mental characteristics with reference to the organism previous to birth and life, we have now to consider the influence that any peculiarities of the organism, especially of the brain, may have in inducing its possessor to choose a particular line of observation or reflection, or to cultivate a special pursuit. In other words, we have to consider what part the organism itself has in determining the Will, and the direction of the faculties.

Having already admitted that the human organism, including its constitutional temperament, with which the nervous system sympathizes, is inherited, and that it owes whatever specialties it may possess at the time of its birth to parentage, it follows that we commence this part of our argument with the unqualified acknowledgment that individual men are subject to great differences of sensitiveness of feeling, and of calibre of mental power.

It must be obvious that whatever may be the character of the organism, and whatever its capabilities, whether special or general, the work done by its possessor in the way of thought or act in the formation of character or regulation of conduct will depend upon the use he makes of his powers. In short, the manner of their employment, and not the mere fact of their possession, is the determinant factor in the result. We have therefore to consider what are the nature and extent of the influence which the possession of certain capabilities exerts in inspiring or determining the use and direction of the powers of their possessor. This is the point on which this part of our discussion turns.

At this stage we are met by the claims of phrenology, which asserts that the volume of the brain, and the relative proportions of its different parts, determine to a greater or less extent both the character and the conduct of every individual. If it were necessary to our argument either to substantiate or to refute the validity of these claims, the task would exceed the compass of these pages. To demonstrate either position would require a very great mass of evidence which, if even we had it at hand, would necessarily be so complex, so conflicting, and so inextricably involved with factors outside of the phrenological data, that the issue would be as doubtful in its nature as it would be difficult to estimate. We were much impressed in our youth with the promise which phrenology seemed to hold out of becoming a powerful aid to mental science, and we have never since ceased to feel deeply interested in its progress. This, however, has not hitherto realized our early expectations.

What we now propose, therefore, is to advance from the range of our own personal knowledge such facts and arguments as tend to show the influences that practically decide men in the employment and direction of their individual powers. We cite them as facts quite irrespective of the bearings they may have upon any particular theories, and we leave the application of them to whatsoever use they may serve for others, their use to us being as evidence of the *principles which actuate human effort.*

Before commencing this part of our task it will be convenient to allude to the remarkable effects often produced on individual conduct by the most trivial circumstances. The fact is of great interest in itself, in its bearing upon personal and mental history, as showing in what slight circumstances the greatest events often originate. As the little switch diverts a mighty engine and its dependent train from one line to another, in widely different direction, so, in like manner, all the forces of the individual nature are often diverted from a course they were pursuing energetically into another entirely divergent therefrom. A little insignificant-seeming switch of Thought, word, or act, is thus proved by daily experience to be able to entirely divert the whole powers of the individual nature, and to give them a new and widely different direction. However powerful in themselves the faculties may be, and however energetically they may be acting, they all obey the little switch, and leaving the line they were intently pursuing, instantly proceed in the new direction, with as much energy as was manifest in the previous course. And if in cases which become conspicuous by means of the important results so arising, the force of trivial diverting causes is recognized, how much more general must such influences be that escape observation on account either of the unimportance of the results or the obscurity of the diverting cause. Indeed, a little reflection would satisfy most men that the operation of the switch is very general in great and small matters, concerning both individuals and communities. It will also appear certain that if the actuating principle of conduct consisted of a fixed part of the internal organism, it could not be suddenly, permanently, and diametrically changed by a trivial external circumstance having no relation thereto.

It seems a hopeless task to reduce to rule forces the operation of which is proved by daily experience to be subject to such change of direction, so sudden and complete, by causes sometimes so trivial. The future direction of such forces could not possibly be predicted by means of any observation of their present action, nor by any estimate of probabilities. They seem to be altogether beyond the reach of ordinary rules or calculations, more especially as the diverting circumstances are generally external to the individual actuated, and of an accidental or casual nature. Nor is the sphere of the operation of the switch restricted to any particular department of Thought or conduct. A sinister *look* may arouse a suspicion the effect of which may never be eradicated and may operate in many and powerful ways. A profession or pursuit originally adopted by choice, and followed for many years with success, or even with distinction, has within the range of our knowledge been suddenly changed for the rest of a life by an incidental remark of the most trivial nature, without any particular present significance or future bearing. In this case the new direction of Thought and effort involved an entire change of occupation, and the exercise of abilities of a different order from those of the previous pursuit. The operation of such diverting circumstances must be familiar to almost every individual. In love, in war, in youth, and old age, amongst rich and poor, such causes are in continual operation. In so far at least as is indicated by such changes, it does not appear that either the tenor of life, or the direction of the faculties, is dictated, or materially influenced, by the special mental capabilities of the individual. Nor does it appear that these offer any kind of opposition to the Will when so actuated.

It follows that if, for example, in the case last mentioned, the first pursuit had been dictated by the force of preponderant special capabilities, that same force must afterwards have actuated the individual in an entirely different direction, involving quite different specialties, and must have undone its own previous work! If, on the other hand, the first pursuit was not so determined, then special capabilities had nothing to do in the matter at all. Both alternatives oppose the doctrine (so far at least as all such cases are concerned) that individual action is decided, or materially influenced, by a preponderant capability of a special character. We therefore conclude that there was a power of the mental constitution superior to that of special cerebral capability, and that this power decided the conduct in the cases above described.

(2) Men do not necessarily employ all the powers they actually possess, nor is the employment of the powers generally regulated by their relative capabilities. This is a fair inference from the foregoing argument, but is otherwise proved by experience. The possession of great physical strength does not *impel* its possessor to use it, otherwise than to serve his general purposes, for which its remarkable capabilities may not be at all required or used. A person so endowed could bring it to bear on any occasion, if he were *otherwise* so impelled, but the power itself exerts no impulse. The man of prodigious strength is not obliged by reason of its possession to exert it, nor even to choose a pursuit which would call its special power into use. The possession of very remarkable powers of vision might induce a man to choose a particular profession or pursuit in which it would give him advantage, but in that event the choice would be an act of the judgment and Will, not an impulse of the *eye* itself!

No one would conclude on seeing a man of great physical proportions that therefore his physique must necessarily have impelled him to devote himself to athletic exercies. On the contrary, it would not be at all surprising to learn that he was a professor of music or mathematics. The influence of special capabilities is not claimed by any one on behalf of other limbs than those of the brain itself. Why then should they be regarded as having any special power of influencing the Will other than that of the rest of the organism to which volition extends? It might indeed be pleaded by a thief at the bar that he was impelled by a strong bump of acquisitiveness to commit the theft charged against him, but the court, acting on the principles of common sense, based on the experience of mankind, would repudiate the plea, and make the thief responsible for the power of volition which every man possesses, and which is superior to all such pretended impulses of the organism. Such impulses are not regarded by mankind in general to have prevailing force as actuating principles.

Men generally are not, in fact, conscious of the relative power of their mental capabilities, and could not therefore be reasonably expected to be actuated thereby. And where such consciousness of special capabilities is supposed to exist it frequently happens that they are not really present in any special force. We have in mind the case of a youth who was supposed, both by himself and his parents, to possess such remarkable aptitude for engineering that we were consulted as to the best course for him to cultivate his special talent. A careful discussion with the youth himself of some inventions he had in mind proved conclusively that he was sadly deficient in the qualities of which he and his parents seemed so proud. It is only just to mention that he kept the secrets of his inventions from his parents, and indeed seemed nervous to disclose them to us, though we were pledged not to divulge them. In this case the prompting, actuating influence in the youth's choice of profession, based upon his own supposed powers, could not have been due to preponderant special capability.

Another case of a different type occurs to us.

The teachers of a child of twelve, who were persons of great experience, reported to his parents that he had no talent for music, that he was making no progress, and had better give it up in favour of drawing, for which he had great ability. This advice was given after two years of trial. The parents declined to follow the advice proffered, believing that application only was necessary on the part of the youth. A special inducement was held out to him as a stimulus to exertion, and within two years afterwards very remarkable ability was proved by his performing high-class music with more than usual skill. The special gift in this case came with application. It follows that the real ability existed, but so far from influencing its owner he was unconscious of it, to such an extent as not to believe he had it. It escaped the notice of both teachers and pupil. Our experience would supply numerous instances, of which the two cited are typical. We may mention that, in the case last mentioned, further experience proved that the supposed superior ability for drawing was also mistaken, for the pupil's ambition that way led to long and laborious efforts, which were not rewarded with nearly so great success as that achieved by the power so much disparaged.

One other case, also typical of a common class, may be cited. A man of our acquaintance possessed extraordinary powers of performing on the violin. He could have earned a better living by that accomplishment than by the pursuit in which he was engaged; but he thought little of his real powers, and disparaged them in comparison of others in which every one, except himself, could perceive his positive weakness. Still he plumed himself on this and set no store by the other.

The illustrations we have cited are typical and represent a very considerable mass of evidence afforded by universal experience, to show that in so far as such cases are concerned the ascertained capabilities have not been the actuating principle in determining conduct. They all indicate the existence of a power or motive force superior to the special mental capabilities. We feel at a loss to estimate, or even to understand, the actuating force of a power of which the possessor himself is unconscious, and about which he may be, and sometimes is, entirely mistaken. On the other hand, every man is conscious of a power which he can and does continually exert; one which may be stimulated even by slight causes to determine conduct, and to divert action from any given direction to almost any other.

3. The evidence afforded by extraordinary mental attainments does not appear to be referable to special talents as distinguished from general capabilities. Indeed, the occasional occurrence of such exceptional cases appears to be a natural consequence of the want of systematic culture generally in early life.

The doctrine of special or inherited gifts, as a means of accounting for exceptional attainments, for which no apparent adequate means seem to exist, proceeds we think upon mistaken premisses. Such attainments being in all cases the result of several factors, and not of some one in particular, the concurrence of the several factors, all necessary to the particular attainment, indicate general rather than special capabilities.

Mozart, for example, is a conspicuous instance of what is commonly called a "born" musician. Let us then consider what is thereby implied. He was undeniably a wonderful master of music, and became so at a remarkably early age. He was even then a fine performer on one or more musical instruments, and composed original and artistic works. For these accomplishments several different kinds of ability were absolutely necessary. For the execution of his music he must have required great dexterity of hand, which could not have been attained without considerable effort and practice. A fine appreciation of time and rhythm were equally essential to his musical execution. For the composition of his original work, he must have possessed a thorough knowledge of harmony, as well as fine conceptions of melody. For the notation or transcription of his compositions, he must have understood the grammar of music, counterpoint, and the art of writing. In addition to these essential elements or factors of the ultimate result, he must also have had, as indeed he actually had, opportunities of considerable extent, and of a very available kind. Of all these factors, equally essential, though in various degree perhaps to the general result, to which is to be ascribed the honour or accident of being the special gift which marked him from his birth as a great, a born musician? If any one of them be singled out from the rest to claim the merit, still the development, the manifestation, the very use of that one would depend upon all the rest, and would therefore be incomplete and inoperative, as a special gift in the popular (and disputed) sense of the term. If, on the other hand, all or several of these necessary factors be regarded as special gifts, the specialty disappears, and their possessor by their means might have been as accomplished in several other ways had those gifts been turned in other directions than that one to which Mozart devoted his varied powers.

Assuming, merely for the sake of argument, that all the qualities above enumerated were conferred in special degree by the inherited structure of the brain, it would still be doubtful which of these diverse qualities determined Mozart's choice in favour of music for employing them, for all, except harmony, would have been equally available for several other attainments. That one only has special reference to music. To us it seems that the convergence of those powers upon the particular pursuit was not due to any one of them individually, nor to the whole collectively, but to the action of the Will. We deem it certain that the Will dominated the mental faculties, and not they the Will.

Shakespeare was another of what are commonly called "born" geniuses. He became a giant in attainments, and has immortalized himself by his works. Still, intimately as these are known, it would be impossible to determine which of the numerous powers they display could have been the special gift that made him great. It would not be quite easy to even enumerate the special powers and elements of knowledge which his works display. His insight into the human mind and character, his intimate acquaintance with external Nature, his wonderful power of diction (with all that is involved therein), his power of discrimination, taste, imagination, and his consummate skill in the execution of his work, all essential to the general result, vie with each other for pre-eminence. Such a catalogue of attainments seems to preclude the possibility of any form of specialty, and to be referable only to *general* power applied with marvellous industry and perseverance.

It is not possible to know how many other men may have lived and died in possession of similar natural powers, but we apprehend that few indeed of those who had the powers had also the energy, industry, and perseverance to make so effective use of them. In this *use* of his faculties, and not the mere possession of them, was the real greatness. Many men possess the necessary faculties, but he had something more than capabilities, he had the Will and the wisdom to use them industriously, qualities far more rare than mere capabilities. In the busy panorama of life he had his senses keenly alert to what was passing around him, noted it well, judged it accurately, and used the knowledge he so acquired wisely and well. He exerted his faculties on the material his environment afforded him for observation, contemplation, and use. He employed his powers with the concentration which effective control alone can give. Industry, when well directed, would make many men great, but gifts so called, rarely, we apprehend, have conferred greatness upon any.

What a flood of light would be thrown upon this subject of actuating principle of mental effort if we knew what Newton was thinking about when the current of his Thoughts was diverted by the fall of the apple, and if we knew also what his subsequent career would have been if that trivial event had not enlisted all the powers of his nature in the solution of the one problem suggested by it. The forces elicited by that simple incident, of a single second of time, in order to solve the problem it suggested, were such as no gift could have supplied, for they created new sources of knowledge and power. The range of mathematical science had to be extended to meet the exigencies of the case. Existing optical means of observation had to be enlarged, new instruments had to be constructed for the gigantic task, and for all these varied requirements a degree of industry, patience, and perseverance were necessary such as would have overcome any but a most determined character possessed of indomitable Will.

Such cases as those we have cited show that the actuating principle, which is throughout this chapter the subject of our contention, is not in capabilities but in the Will. These cases go far to prove that greatness is within the reach of many men if they would devote the powers they possess with industry and perseverance; that it is vain, foolish, and unprofitable to blame inheritance. A supposed deficiency of capability is often a serious source of discouragement, and a feigned deficiency is a still more frequent excuse for supineness. We have in mind a youth who was sent to be instructed in coffee planting. After a few days he complained that he could not climb the hills, being unequal to such exertion. He declared it was quite impossible. He was therefore let off duty, and his friends were applied to for instructions as to his future career. In the interval, he betook himself to his gun. In the eagerness of his sport he went to the low country, 1000 feet below the estate, and repeated journeys thither on foot nearly every day, until it was discovered that he was so employing his powers. He was then made to understand that if he could make such journeys for his sport, he could more easily go the easier distances to his work.

Many instances might be adduced from the sphere of our observation of great attainments being made by persons of ordinary ability by resolute effort. Quite lately a youth, who had no knowledge of Greek, desired to matriculate at the London University. He was fairly advanced in other respects, but of Greek he did not even know the alphabet. However, he had just three months for the task of qualifying, and he did it successfully and passed. A missionary of our acquaintance acquired a sufficient knowledge of the Portuguese language in six weeks to be able to compose and preach to a congregation of that nationality. And we have reason to know that he did it very creditably. The eight known books of Euclid were mastered thoroughly, and all the exercises at the end of each were answered, by a youth of our acquaintance in two months. This work was not done under pressure as to time. The only stimulus was an earnest desire on his part, to master the subject; moreover, on completing his task he was not conscious of having accomplished a feat. The case previously mentioned, in which musical talent was supposed to be wanting, was another instance of the power of Will, for the progress was very rapid after the effort to learn became earnest.

From such cases, which are by no means uncommon, it seems that the doctrine of inherited specialties, as a means of accounting for exceptional attainments, is unnecessary, and that to all the cases typified by our illustrations it is inapplicable.

The foregoing remarks on special capabilities, as

distinguished from general power in the physical organ, are equally applicable to other mental characteristics. The argument and its grounds are the same in all. In some cases, such for instance as a habit of lying, example generally affords a sufficient means of accounting for the characteristic. We know of parents who see this terrible propensity in their children, and are filled with shame and distress at what they are obliged to acknowledge. Still, they are sure the habit cannot have been contracted from themselves, and they therefore refer it to some distant deceased ancestor. But this resort is unreasonable and unnecessary. There are much more cogent causes always at hand. The habit of falsehood in children is almost always contracted by example in the shape of what are called harmless deceits. These are practised by many parents and tutors, in one form or another, and children do not and cannot discriminate between the so-called harmless and the more mischievous forms of falsehood. The example is the same in both, and in fact there is no real distinction; moreover, parents are not a child's only exemplars, and the propensity to imitate, as well as the power of discernment, in young children is very remarkable. Hence few children escape learning falsehood, and fewer still avoid the practice altogether. The only case in the range of our knowledge was that of a child who was sedulously taught by the example of parents who thoroughly believed in the real nature of the infection. Being an only child, and seldom entrusted to a nurse, the influence of example was vigilantly kept pure. In that family corporeal punishment was practised, though with much moderation. Still, corporeal punishment in any degree is a source of severe temptation to children, and is, even on that account alone, deserving of serious consideration. The child, though subject to this test, was never known during our acquaintance with her, to depart from strict truth even with the fear of punishment in view. Example in her case was effectual, and if she ever learnt to lie, she acquired the accomplishment in later years since we lost sight of her.

Until tutors recognize the actual effect of example, and the actual power of observation and discernment in the youngest and most unconscious-looking children, and until it be practically regarded that the youngest children are *always* learning, example will continue to instil the prolific germs of falsehood and other forms of vice supposed to be harmless. Attempts, sometimes very clumsily executed, to conceal something stealthily from a child's view, will suggest secretiveness, which may become habitual or not, according to circumstances; but when the idea is given, the use that may be made of it is no longer a matter of example of others, but of practice at home. The idea is a property which may either be forgotten and fall out of use or be used habitually.

The fact that in their earliest stages infants are practically regarded and treated as if they were taking no notice, seems to us to account naturally for the development in them of tempers, ideas, and seemingly unaccountable specialties. During that period they are allowed to see and hear many things which they are not then supposed to notice, and they thus gather into their mental stock many other things in the same unobserved manner in which they actually prove that they acquire language. Such being undoubtedly the fact, it must naturally happen sometimes, according to the doctrine of probabilities, that there occur remarkable concurrences of favourable or adverse conditions, such as to account for the appearance and development in them of extraordinary and seemingly unaccountable phenomena of special power, of unexpected bent or direction, of strange temper, and, in short, of any and every form of idiosyncrasy.

Considering, moreover, how powerfully and completely the faculties may be directed into particular channels by trivial circumstances, the work of an instant, and considering also the powerful influence of the natural propensity to imitation, there seems to be no need whatever for recourse to any such doubtful theories as inherited specialties, to account for occurrences natural to the chance conditions of the mental history in early infancy.

4. Irrespective of the foregoing considerations, the doctrine of inherited temper, disposition and *special* gifts is not consistent with the facts of daily experience.

Instances are of common occurrence which show that some men actually do a great deal more effective work with small and imperfect means and appliances than others who have larger means and more ample opportunities and instrumentalities. And this fact is as manifest in regard to brains as it is to other parts of the physical organism. It follows that what is inherited, the organ or instrument, does not in fact determine the use that will be made of it nor yet the power of using it. A powerful brain, therefore, phrenologically powerful, that is, is available only in so far as it is employed, and even then it needs also to be well directed. Who does not know in the circle of his own acquaintance of men possessed of great natural capabilities who can scarcely be induced to employ them? and of others who use such powers for purposes utterly unworthy of themselves, or their great capabilities? Evidence of this nature abounds in every individual experience, and need not therefore be further enforced. It proves that the materialistic theory which makes the organ rule the man is not borne out by, nor at all consistent with, the facts of experience.

The fact that there are often in children great resemblances to their parents is, of course, undeniable; and considering the close relations which subsist between children after birth and their parents, such resemblances are less surprising than the remarkable dissimilarity that generally prevails in families. Considering the powerful influence of the natural and universal propensity of imitation, the wonder is not that there should be strong resemblances, but that they are not more numerous and more striking. Such resemblances appear to us to be even fewer and less obvious than this propensity alone should ensure, quite irrespective of inheritance. The great dissimilarities would offer more difficulty to the student of actuating principles if it were not so well known and so abundantly proved that very slight incidents are capable of setting up very marked lines of divergence from any given course. We fail therefore to see in family resemblances, as regards mental action, any evidence in favour of the doctrine of inherited tempers, states of mind, or special capabilities, though we think the prevalence of dissimilarity is very strong evidence against inherited tendencies.

According to the evidence derivable from inherited qualities in the lower animals, and the uniformity that generally prevails in the relations of parent and offspring, it would be reasonable to expect a considerable degree of regularity in mental qualities if these were really inherited. The fact is, however, that in these respects the utter irregularity that subsists amongst children of the same family, not occasionally but almost universally, seems to be conclusive evidence that the idiosyncrasies of individual members must be derived from different sources and not from one common inheritance. The diversity is easily explicable and natural when it is considered apart from inherited qualities, and is referred to the sufficient natural causes to which allusion has already been made in the foregoing part of this chapter. If other circumstances which are known to influence the development of character suffice to explain both the resemblances and the diversities that exist in the mental characteristics between parents and children, we see no reason to resort for explanation of them to a theory which is of doubtful authority, and opposed to general experience.

5. As man is endowed with natural faculties, wherewith to acquire knowledge and to act for himself, it is not reasonable nor consonant with analogy that his natural powers should be superseded or their functions anticipated by other means.

In all the works of Nature, so far as these are known, there is one universal rule which prevails throughout—namely, that all existences commence their natural history from zero, and thence proceed through successive stages of progress to the maturity of their development. The human mind does not appear to be an exception. The mental history, beginning with life itself, seems, in all its attendant circumstances and conditions, so far as these may be observed, to support the view that mental operations begin in the infant mind like all other known things from zero.

The long term of physical helplessness through which the human babe has to pass in order to acquire the power of using his corporeal faculties independently, in which he differs so much from the offspring of the lower animals, which acquire the use of their faculties so soon, seems to have been designed for the purpose which it subserves so well of giving time and opportunity for those mental acquirements so necessary to the life he is destined to enjoy. The sphere of mental operation for which the babe is to be prepared differs from that of the brute in the assertion of a higher nature over that of the lower, to which the brute is permanently subject. The mental learning of the lower animal is confined within the narrow limits of the teaching of a parent whose entire stock of knowledge is bounded by his own individual experience. Such knowledge, so limited in its sphere, and affecting the lower nature only, is easily imparted, simply imbibed, and as readily applied to its narrow uses. The mental instruction the babe has to acquire, on the other hand, embraces a far wider range, extending beyond the narrow limits of the lower nature into another and higher natural domain. The kind of knowledge the babe has to imbibe is neither restricted in its sphere by the individual experiences of the parent, nor by the limits of the lower nature. The human parent draws his knowledge from the accumulated experience of the race by means of language, and the babe imbibes from his parents and tutors his first ideas from the example which embodies, in certain measure, those large experiences and the principles they involve. Moreover, the human parent derives his inspirations from another and higher domain than that of the lower or animal nature. Moral principles, which distinguish him from the brute, are also embodied in the example of the human parent, and are thus exhibited for the imitation and instruction of his offspring. Hence comes it that the sins of the fathers are visited upon the children from generation to generation, not by any special indignation of an offended Almighty, but by the operation of natural law. And, thank God! mercy and love are also by the same means imparted to thousands, and will prevail in proportion as men recognize the natural means involved in the Divine ordinances, of which the one now under contemplation—that of the infant's opportunity for mental and moral acquirements—is not the least beneficent.

Such, then, being the nature of the mental work to be begun, such the opportunity providentially afforded for the work, and such the agencies and appliances by which it is to be accomplished, is it reasonable to suppose that infants would be launched into life with such mental and moral furniture as would render those conditions superfluous and embarrassing?

Regarding the brain as a part of the material organism, having certain functions, as other parts of the organism have, each for its particular use, and dependent like them upon vital force and Will to employ them, we do not understand, nor can conceive of its acting of itself and upon itself, any more than that a hand or other limb should act irrespective of those forces. It does not appear to be more reasonable that the brain or any of its parts should operate of itself, than that a violin in the hand of its owner should assert itself and play of itself. We have no experience of any such tendency, and we think we could scarcely have been so actuated by mechanical or any form of impulse of the brain, without being conscious of something opposing our free-will from that quarter! Nor does the law of the country, or the common experience of mankind, sanction the idea practically of any such force superior to the Will. God did not mock His creature, man, by imposing upon him obligations which he was not able to fulfil. He who gave the command gave also the power to obey it; and every man, however strongly he may be tempted, is conscious of such power, unless he have lost it by abuse.

The materialistic doctrine, that matter can act upon itself, may not be held theoretically by all who believe in the self-assertion of the organism of the brain, but it does not seem to be really separable from that idea. The several forms of the principle, whether that a piano should play upon itself, without any force external to its mechanism, or that it should compel its owner to play upon it, or that the tunes should exist therein ready made, are all equally inconsistent with experience, inexplicable, and contrary to known natural laws. We conceive that, without some form of force external to itself, the material organ would be inert, as it really becomes when life ceases and was before life supervened. We apprehend that the infant, at the time of birth, has an organ of which the appropriate use is for mental work, and that he has also a motive-force which sets it in motion. The instrument may be of a more or less sensitive nature, of greater or less volume, according to hereditary constitution, but its action will depend on the force of the Will.

Supposing that exact functions could with certainty be ascribed to certain parts of the brain, and supposing it were proved that the proportional development of these several parts to each other and to the whole mass of the brain determined the power and influence of each, still the use of these parts respectively, as of the whole collectively, would depend on a force external to the instrument itself. A part possessing great relative development would enable its possessor to exert more power by its means than by a weaker part. Still, as has already been shown, the efficacy of such part would depend upon its being used appropriately; and we know that men do not necessarily, or even generally, use all the powers they possess, nor do they always use them well or wisely. Men inherit brains, no doubt, and so also eyes and ears connected therewith, but they do not inherit Thought or temper, any more than scenes in the eye or music in the ear.

The evidences afforded by phrenology, in the present state of the science, are in several ways doubtful. It is not at all certain to what extent the development of the whole cerebral mass or of its parts depends upon the use that is made of them. The fact that an unused limb is generally very feeble until habituated to use, and that its further development beyond its normal condition also depends upon its being exercised, suggests the idea that the mass of the brain, and the relative proportions of its parts are in some degree (perhaps very considerably) influenced in both those respects by the measure in which they are employed.

Moreover, the brain itself does not conform in all cases internally to the external appearance and development of the cranium. Hence, these are apt to be misleading. The mode of growth and form of nourishment of the cerebral substance itself are also, though perhaps less important, factors of the general development and power of the brain.

We lay much stress on this part of our subject, because of the important bearing it has upon the training of the mental faculties, and because it shows what may be done by the means which all men, in greater or less degree, possess in common. The danger to be apprehended from the doctrine of inherited gifts, and the dependence upon special capabilities, is in many ways serious. It is a great discouragement to effort to think that it may be unavailing. A prospect of success, and the hopes it inspires, are great incentives to effort; and, on the other hand, the feeling of despair created by a doubt of requisite qualification is correspondingly disheartening.

The belief closes up certain paths which would otherwise be open to a pupil's ambition, or to the choice of a calling. Worse still, it disparages the most powerful agencies, industry and perseverance, as means for attaining excellence, and makes this dependent upon the accident of some special inheritance. It serves as an excuse for idleness, disinclination, and inattention, both to learning and to moral principle and duty. It puts into a false position fellow-workers who have excelled by reason of their industry, and makes them the objects of jealousy for their higher supposed endowments, instead of being admired for their real merits. To set against these evils the theory has not one redeeming feature. The steward of one talent was not therefore excused for the neglect of his restricted trust; and the commendation for the improvement of two talents was the same as that awarded to him who had improved the trust of *five*.

As we write, there comes to our recollection a man who, with a paralytic arm and a lame leg, abandoned a sedentary profession, in which he had good prospects, in order to follow by preference another pursuit, for which of all others his physical infirmities seemed to be the most effectual disqualification. His indomitable spirit recognized in them no impediment. We can feel in imagination the hearty grip of his only capable hand, and see his bright cheery countenance, animated with its habitual sunshine. We recall the view of his crooked figure, and eccentric gait, climbing an Indian Ghaut, at a gradient of one in two, up a rocky path, such as only the khuds of Indian mountains boast. There he pursued his new calling in cultivating coffee, which loves such precipitous ground. Few of his abler-bodied neighbours could outstrip him in a difficult ascent, or in his daily work. His coolies borrowed the light of his countenance, and reflected it in theirs. On the high road he was the life of the company, and when he quitted the rest bungalow, and jerked himself into his saddle, he would start the laugh which his grotesque movement was sure to provoke, in order that his fellows might indulge it without fear of offence. With his principle that though the babe can *do* very little he is capable of suffering a great deal in that trying stage of his history; and further, that what he suffers is far more influential on his mind than anything he himself has power to do. It is proved beyond doubt that during all that trying experience he assuredly does think, and as he has then little else to think about but his personal feelings and his small environment, he will certainly exercise his thinking faculty on the wants and pains, the satisfactions and disappointments, the joys and griefs, the attentions and neglects of which nearly all his experience then consists.

The aim we have in view is to establish a right *beginning* and direction of the use of the thinking faculty, in order to save the consequences of neglect, which involve serious trouble to tutors, and grief to their pupils. We desire by our methods to prevent harm from accruing, in order to save the necessity for a great deal of painful undoing afterwards.

Our methods are in principle contained in the wholesome old rules laid down in Scripture, and applicable to all periods of life, and to all sorts and conditions of men—viz., to *do as we would be done by; to sow as we would reap;* and to *overcome evil with good.*

To put these universal principles into appropriate form for our particular purpose we have to consider the conditions, bodily and mental, of those to whom they have to be adapted. As regards the babe, these conditions, as laid down in the preceding chapters, are as follows:—

That from his birth the babe has very tender susceptibilities and pressing wants, which at first absorb his attention.

That his dawning intelligence has then no other sphere for its operation than is afforded by his wants and feelings. In them it takes its rise.

That he is not unconscious or indifferent to what goes on within and about him, but, on the contrary, his Thoughts are occupied therewith continually, and his emotions will necessarily be such as his experience of them induces.

That his first relations with the world outside of himself are thus commenced.

That his utter helplessness makes him entirely dependent upon his tutors for the supply of his wants, and for the regard due to his tender susceptibilities.

That occasional neglects of his wants and disregard of his feelings will be far more influential on his mind than frequent caresses and fitful indulgences.

These conditions afford the best possible opportunity and means of enabling tutors to acquire and maintain complete influence over him, to attract his love and confidence, and to thus win his willing and natural obedience. The measure of the babe's necessities, on the one hand, and of his sensitiveness and helplessness on the other hand, define the measure of the power and influence placed in the hands of his tutors for that very purpose. Their use or neglect of the influence thus afforded determines the nature of the issue or state of mind produced.

The foregoing conditions determine the nature of the first inscriptions made on the blank tablet of the infant mind, and those conditions, as well as the other considerations respecting baby life contained in our preceding chapters, suggest the following adaptations of the three main guiding principles already mentioned as rules for the training of the incipient faculty.

1. That the babe should be regarded and treated with due consideration of the facts that he is ever conscious and observant, and that his first ideas and emotions, being the natural product of his experiences, will necessarily accord therewith.

2. That tutors should regulate their conduct towards him and their example generally with special reference to making the best use of the divinely appointed means and opportunities which the conditions of baby life afford, for initiating in his mind ideas and emotions of love and trust, the cardinal principles of right-thinking and of Christian virtue.

3. That the earliest indications of Will should be carefully watched and delicately treated in such manner as to mould it to voluntary obedience.

4. That moral principles should be exemplified practically in all nursery proceedings, especially in regard to exhibitions of temper, and in strict avoidance of every form of guile, deceit or secretiveness, however harmless they may *seem*.

5. That shocks, and all means of creating terror or exciting sudden emotion, should be sedulously avoided.

We now proceed to make a few remarks on the application of the foregoing rules.

1. If the responsibility which the conditions of baby life actually involve were fully recognized, and if the fact of the observant power and habits of infants were duly felt, there would be little if any need for any other rules than would naturally be thereby suggested. Hence, at the outset, we would crave the careful attention of the reader to the sum-

mary of the mental work which actually must have been done during babyhood, as it is described at pp. 109, 110.

We fully admit that great care and tenderness prevail in current nursery practice, yet, with all the regard so manifested for the sensitiveness and helplessness of the babe, there is, nevertheless, mingled with it a great deal of inconsistent disregard of both his feelings and his wants. Such lapses, however, are far more influential than the attentions he receives, and indeed these make him all the more sensible of them. Baby is not peculiar in being much more sensible of neglect than of attention, of inconvenience and disappointment than of ease and satisfaction. This is a characteristic of the race, irrespective of age. Men think comparatively little of their comforts and blessings until they lose them, and, even then, they think rather of the suffering which want entails than of the enjoyment which satisfaction procured. Hence, it should be borne in mind, that the little provocations, neglects and discomforts the babe suffers, are much more effective items of the general account than the fitful caresses, or even than the habitual consideration of which they are frequent accompaniments.

Here is a scrap of every-day experience. As we were lately occupying a bedroom adjoining a nursery, we heard, as we lay awake one morning, that baby had awoke and was cooing and crowing. This went on for some time, and the tone became gradually complaining, and at last loud enough to awaken nurse, who muttered a few words, and all was silent for awhile. Again baby resumed his complaint, which became stronger and louder by degrees, until at last, having quite exhausted his patience, he burst into an angry cry. This was promptly hushed by some form of repressive measure to stifle the resentment neglect had kindled. This is a trivial incident, too trivial some persons may think to be worth mention. But we adduce it as an item in the natural history of temper. It is by such means that the worst of tempers are wrought up in the infant mind to blight the happiness of several lives, not of his only whose the temper is, but of those who have to endure it.

As a single occurrence, we do not pretend that the incident just mentioned would permanently influence the feelings and temper of the child, but, as an exponent of what he may experience daily or perhaps hourly, it is the key to the habitual state of his feeling.

Resentment is itself a bad feeling. Begotten of evil by neglect of duty, or by provocation, disregard of feeling, or by any other means, it begets evil. Its operation in adults is qualified by experience and self-control, and by such allowances as circumstances may suggest; but the babe has none of these resources to modify his anger. Is he less sensitive on that account? or will the effect of irritants be less in him than it was in those lower animals to which allusion has been made? or is it of less moment to create an ill-temper in a human infant than in a horse or a monkey?

2. Wants and appetites are the springs of a large proportion of the feelings and thoughts of baby life. Hence they supply abundant opportunities and means, by judicious ministrations and satisfactions, of inspiring his confidence and love. The sweet tones of a mother soothing her babe at the breast

and caressing him whilst she is satisfying his natural craving, add joy to his contentment. The occasion affords priceless opportunity for inspiring love and trust as well as satisfaction. And it is also a most effective means of acquiring that subtle power over the babe which prevails without effort on the one part or sensible constraint on the other.

Seeing that baby-Thought must sympathize with the experiences that suggest it, how much his state of mind must depend on whether, for example, he be reclining at ease in the arms of his nurse, or be huddled up comfortlessly in an awkward heap; whether lying peacefully, listening to a sweet lullaby, or be scared by howls to drown his senses and force him to sleep; whether he be gently soothed, or shaken and tossed into a state of stupor and bewilderment. In all these, and a hundred other such alternative situations, his feelings and thoughts will necessarily be induced by, and will correspond with, the originating cause. It could not be otherwise, for effects depend upon their causes as much in mind as in matter, in babyhood as in later life. Whether, therefore, the end sought be to establish a happy relation with the outer world generally, or to inspire confidence and love towards tutors in particular, the recourse is ever the same.

3. The Will must be moulded by insensible influence, never forcibly repressed, nor impatiently coerced. However kindly it may be employed, coercion excites resistance and produces resentment in some form. This feeling may or may not be at once manifest, but its results on the mind are essentially evil. It may not lead to any active demonstration, but *thought* and *feeling* are always

active *within*, whether visible externally or kept secret in the bosom. The idea that because babies do not always manifest resistance they do not therefore *know* or *feel* it, is a fruitful source of the ill-tempers and headstrong stubbornness of spoiled children. Unless therefore they could be made insensible of being hurt, provoked, and deceived, and capable only of feeling kindness, satisfaction, and happy influences, unless, in fact, they could be made to remember and to be influenced by no other feelings than those of contentment and joy, the treatment they receive must necessarily produce its *natural* results in their minds. They will be wilful or obedient, resentful or amiable, according to the prevailing feelings induced by their experience. Of course, natural temperament will make them more or less sensitive, and affect the *degree*, but the *kind* of feeling will be the same in all.

4. Precept being of no avail until language has been acquired, the only means of inspiring ideas of moral principle, and of educating the conscience previously to their acquisition is example, for which nursery life and the conditions of babyhood afford ample scope. Seeing that children are always observant, with more or less attention, of all that goes on in their restricted environment, the influence of example is always operating and suggesting ideas of one kind or another. Moreover, we have already pointed out the fact that the ideas and knowledge in general, which children derive from observation and inference, constitute the great proportion of all they learn, and they are therefore always *learning*.

5. This refers to a class of nursery practices of a highly reprehensible and cruel nature, which are

often productive of life-long results, such as antipathies, extreme nervousness, stammering, epilepsy, and even idiotcy. No sensible nurse or parent would justify recourse to such means theoretically, but almost all practise them in one form or another, especially in the period of infancy, without thinking of consequences. Practically, consequences are too little considered in regard to baby Thought. It is more convenient to assume that babes do not think or will not remember.

IN INFANCY.—When baby can walk and begins to try to talk his attention is not so much engrossed with his feelings as it was at first. He is no longer a passive observer, but is anxious to use his acquired powers. His dominant impulse, *activity*, is a characteristic of youth generally in the brute as in the human race. The lamb, the calf, and the kitten are all exuberant in their particular ways, and frolic with an overflow of joy with which the manifestations of infant mirth are not to be compared. It may be that they have no proprieties to observe or to learn, no fashions or formalities to regard, and but few restraints to suffer. Their toilet is of the simplest, rather a pastime than a penance. The young brute has no soap in his eyes. His limbs are not thrust into garments for appearance' sake, nor does he suffer the constraints and chidings which form so irritating a part of infant experience.

Whatsoever tends to deprive children of the innocent joys proper to juvenile life defrauds them of a part of their natural birthright, and contravenes the order of Providence. A sad, dejected-looking child is an anomaly, an object of our deep sympathy, not for its present indications only, but especially for its future effect in beclouding the aspects of the future life. The relation of an individual towards his fellow-men is a principal factor in the sum of life's experiences, and is the product in general of the treatment he has received, or of the manner in which he has learned to regard it. The effect is seen in those who brood over their disappointments and vexations, instead of letting their first experience of them suffice. Gloomy thought may not be so manifest in its effects on the juvenile mind but its natural results are the same. The moral influence of a joyous youth may be best inferred from the effect produced on the character by an opposite state of mind, as may be seen manifested in the moroseness and habitual discontent of disappointed and thwarted men. It is rare indeed for such characters to exemplify any of the nobler qualities of virtue.

The rules already prescribed are applicable to these and to all future stages of life. Corporeal feelings and mental susceptibilities are not always so sensitive in after-life, nor do they exercise the dominant influence they possess in babyhood, but they are always entitled to consideration and respect. They cannot be disregarded at any time, without adding to the painful side of experience. The treatment men receive never loses its efficacy as a means of inspiring confidence and love, or of exciting distrust and indifference. Nor does example ever cease to be influential, though it is not so effective in after-life as in its earliest stages.

Coercion in every form is to be avoided, as it is destructive of the best chances of success and is otherwise evil. In whatsoever degree coercive measures may seem to be successful, they produce only the semblance of obedience far different from and inimical to the true principle.

The general method we recommend for all the proceedings of these periods for the instruction of the juvenile mind is one of *correction*. This term, as commonly used, unfortunately carries with it the idea of punishment, which is the opposite of what we propose. Punishment in every form which involves personal humiliation or corporeal inflictions is wholly inapplicable to the period of infancy, and is of very questionable expediency in any stage of youth. It seems to us to be opposed in principle to the scripture injunction to overcome evil with *good*. As a corrective, punishment is of very doubtful efficacy. It is at the best a deterrent, and even in this respect it is rather suggestive of avoiding discovery than of repressing evil at the core. In our experience the effect of corporeal punishment is to harden and inflame the feelings, and therefore to multiple evil instead of discouraging and extirpating it. It does not affect the *motive* but represses action without correcting, in the sense of reforming, the character or affecting the heart. Nevertheless it is often professedly based upon the example of God's government of the world.

With all the respect due to so great and learned a divine as Bishop Butler, we cannot agree with the view propounded in his immortal "Analogy," that God's government of the world is one of rewards and punishments in the proper sense of the terms. It is undoubtedly true that the choice of the *good* and the eschewing of the *evil* yield the reward of virtue, and that an opposite course leads to the wages of sin. But the benefit in the one case, and the suffering in the other, are not visitations of the Almighty, but the natural results of the operations of good and evil. When Adam disobeyed the command the penalty was not inflicted upon him, but was worked out by him. Death was not visited on him, but followed after a lapse of time by the operation of natural law. Death was not a punishment, in the proper sense of the term, but the natural consequence of sin, as is shown by the fact that it was so long deferred. Moreover, the death of Abel was no punishment, for it fell upon the innocent by the sin of the guilty brother. The natural effects of sin have been manifest in the experience of mankind, and are to be seen in all their appalling aspects, wherever men congregate. The agency by which evil is overcome is equally manifest in the life and death of our blessed Redeemer, in which the glorious triumph of the power of good over evil attained its highest manifestation. It is also exemplified in the lives of those who eschew evil.

In the dispensations of Providence, evil is left to work out its own results in its own way, and if these had not been manifested in the unspeakable horrors so often witnessed in the world, they could not have been conceived. Man in this life learns by awful experience the law of evil and death on the one hand, and of good and its resultant happiness on the other hand, by similar means to those which teach him the law of light and other of the forces of Nature. All laws are learned by their operations and results, and the laws of good and evil are no exception to the general rule, but are also taught by the experience of mankind. Hence, the respective results of good and evil are not interventions of

Almighty power, but the natural operation of law. It follows that punishment, as such, is not sanctioned by the example of God's government. Nor is it proved by experience to be effective, otherwise than as a deterrent. Of hell we know nothing positive, as the language of Scripture respecting it is figurative, fire being impossible in a spiritual state. What reason and analogy confidently teach is that it must be a state in which individuality will be perpetuated. Any such change as would destroy that would create another being. Man will therefore necessarily there continue to be what he was in this life, except that having made his final election between the rival principles of good and evil which were set before him in this world, he will abide eternally in the condition he has chosen.

For its deterrent influence punishment is inflicted by human laws, and penalties are thus imposed for wrongs committed against society, but these measures have long been found wanting in corrective force. Hence, when a reform of character is contemplated, other means are employed in which evil is met by its natural counter-agent—good. This, therefore, is the agency we would invoke in the training of children.

Correction, in our sense of the term, or the rectifying of what is not right, is the most effective means of educating and training the mind to right thinking, and the conduct in right acting. Our sixth rule therefore is: "*To instruct children in right thinking and in good conduct by correcting errors and defects whenever they arise.*"

The form of correction we recommend consists in aiding, supplementing, and setting right whatever is said or done, in which help in any of these forms may be required, and is not to be done in a fault-finding or interfering way, but with tact and good humour. Such correction has a good influence in other ways beside the immediate occasion, for it tends to confirm the child's attention when it is already engaged, and also manifests interest in his ideas and efforts. Every form of correction has ultimate as well as immediate use. Every mistake, misjudgment, and error of whatever kind affords an opportunity for a lesson of lasting use, administered just when it is likely to be impressive. The principles on which opinions and inferences are formed can never be so effectively explained as by practical application, and so with other mental operations, the principles of which are simple when exemplified, but are extremely hard to formulate into verbal rules. Even when so formulated the difficulty of applying them still remains. It is manifestly, therefore, much easier to learn the principle by practical example than to reverse the process and to reduce a verbal formula to its practical application. Our system of training is based upon experience of the fact that practical lessons are more intelligible and more applicable than theoretical ones, and that principles are better deduced from practical application and use than from direct verbal formulæ, even when these can be clearly grasped. Moreover, the principle of correction applies universally to every form of mental effort and to every kind of knowledge as well as to conduct.

Tact and judgment would rarely ever be at fault if tutors rightly understood the nature and purpose of the work to be done, either in this or any other department of infant training. Hence, our object is rather to explain principles of action than to lay down specific rules. Indeed, these could not be so framed as to meet circumstances so variable and diverse as those of infant life and environment.

The activities of all kinds which characterize this period when infants begin to employ their powers with some independence, are generally *imitative*, but in the absence of experience they are naturally applied without discrimination, and therefore often mischievously, in such manner as to bring children into collision with tutors. This is therefore specially a time for the application of our rule—a time for constant and judicious *correction*. Thus turned to account it affords opportunity to impart to the infant much useful and elementary experience, or, on the other hand, if he be either permitted to do what he pleases or constantly thwarted, bad habits or bad tempers must naturally and surely result. If left to himself his inexperience and ignorance will surely lead him into mischief, for which he may be unjustly punished, or which he may repeat. If continually thwarted his temper will be tried severely, for it is provoking at any and every period of life to be interfered with, and especially without apparent reason. Correction is the only safe method of treatment at this age. Remonstrance and explanation are of little avail until reasons can be readily understood. Till then reasons have to be inferred from the tone and accompanying action of tutors.

In the Appendix (Note 3) we give a typical instance of a bit of unintentional mischief done by a simple act of imitation. The actor in this case was a Newfoundland dog. He was unintentionally instructed in *theft*, and proceeded by imitation to get his master into trouble. We have witnessed many cases in which children, who acting under similar imitative impulse and with the same innocent intention were unjustly punished as for a fault.

There is probably no moral principle more generally or earlier imbibed by children than the idea of justice. There is certainly none which needs to be more vigilantly practised in the nursery. For instance, a little fellow who was just beginning to talk was applauded for a successful imitation of his father's customary exclamation, "Hallo!" Soon afterwards he made an equally successful imitation of another emphatic interjection, for which he was promptly punished. The inconsistency and injustice of such treatment must be an embarrassing experience to a young child, who is imbibing his first lessons of moral principle, by the *only* means he has of learning.

It is manifest that children should not be left to persist in erroneous or awkward ways until some serious consequences ensue to compel a correction. Inexperience and inexpertness will certainly lead to accidents; indeed, some of the most useful of early experiences are so learned. In all such accidents careful discrimination is necessary between the results of ignorance and those of disobedience, or which involve infractions of moral principle. The treatment of misfortunes as though they were faults is a frequent source of trouble to children, who are very sensible of the practical injustice involved therein.

We have already shown in previous chapters that occupation is eminently useful for giving an innocent direction to activities, but it has besides a beneficial tendency in fixing attention, and thus forming a habit which will be of great value to both pupil and teacher in the next stage of the mental progress when tuition comes into use. Occupation, whether of body or mind, is esteemed a pleasure or a task, quite irrespective of the amount of effort involved, solely according as it is done by choice or imposition. Activity and effort are natural and, in themselves, agreeable, but they become irksome when required to be done reluctantly. Adults are conscious of duty and necessity as motives for overcoming their dislikes and directing their activities, but infants have no such ideas to reconcile them to master unwillingness, wherefore occupation should be made attractive to them.

IN CHILDHOOD.—In order to apply our rules to the third period of our natural history, it is only necessary to consider the further progress made by the child in mental development, and the altered conditions consequent thereon. The principles hold good for all periods of life. The inquisitiveness which characterizes this period affords excellent opportunities for imparting knowledge, just when there is an appetite to receive and inclination to digest it. Still, the mental powers are weak, and must not be strained. Opportunities must not be lost, but neither may they be used with good effect beyond the small natural capacities of the pupil. *One thing at a time*, and that one thing simple and clear, should be the rule in regard to all ideas or explanations involving novelty. No one thinks of beginning to teach reading with whole words; letters come first, and words by degrees, afterwards. A long course of spelling words, letter by letter, precedes the power of grasping them entire. In like manner should all new ideas be imparted gradually by means of their elementary components. But this is not sufficiently considered. Explanations are often made, or ideas imparted, which involve matters either new or imperfectly understood without sufficient care as to whether the words employed or the connected ideas are apprehended. Children are expected to grasp them at once, and to follow the glib tongue of a teacher, who, knowing them all intimately himself, is apt to assume that they must be clear to the pupil. Any difficulty the child may manifest is attributed to his want of attention or to natural stupidity. The lady who tried to teach her child to tell the time by the clock (Note 4, Appendix), did not perceive that her explanations were like reading words of two or three syllables to a child who was yet obliged to *spell* words of only one syllable.

At this age children are usually very communicative. They turn themselves inside out with a candour and freedom which show the working of their minds and the nature of their thoughts. They thus afford innumerable opportunities for correction of many kinds, such as the use of words in their proper sense, the inferences they have drawn, the ideas they have formed, and the principles they have imbibed. The lessons given in this manner suggested by such opportunities, and judiciously restricted to the particular point in question are effective, and divested of those elements which so often make lessons disagreeable. A lesson when not suggested by such opportunity always exacts *attention*, but when it springs out of, and responds to the child's own impulse, it finds the mind already attent. Other lessons require suitable opportunity, but these find it ready made. Other lessons may, and generally do, relate to matters of no present interest to the learner, but these concern the thing uppermost in his mind.

Still lessons cannot *always* be so happily timed. During the period now under discussion lessons of a systematic nature must be begun. Hence, they should be gradually introduced when attention can be commanded without strain, and when a sufficient knowledge of language has been acquired to enable the learner to follow *readily* the explanations given. Let teachers bear in mind the difficulty they themselves experience to catch a strange word of their own, or of a foreign language, and they will then understand, that until words are *familiar* to children, they are not instantly grasped. Seeing how important an element of education lessons will become in the next stage of the mental progress, school days, great care should be observed during childhood to avoid their becoming distasteful tasks.

We have already alluded to the fact that children learn a very large proportion of all they know, including the great work of acquiring language, without any teaching. Learning and teaching are not natural counterparts one of the other, for a great deal of the effort usually expended in teaching is barren of beneficial effect, and is sometimes productive of irritation and positive harm. It is important to our present purpose to make our meaning well understood. In all that kind of learning which is acquired by observation and inference, without teaching, the attention requisite for effective formulation, has not to be *exacted*, but is captivated or attracted. On the other hand, a teacher needs to *exact* attention, failing which, his efforts must necessarily be fruitless. Attention is a fundamental condition of learning. Hence, there is a cardinal difference between the efforts required respectively for observing things that attract notice and for attending to things imposed upon the mind. In the former case, the attention, the first essential, is engaged in advance without effort. In the latter, it has to be obtained. In the former case, success is assured. In the latter, it is very doubtful. In the former case, the mind is free from all external embarrassments. In the latter, the eyes of an expectant teacher are upon the learner, whose mind, therefore, is distracted by influences external to itself, and at best these are more or less disturbing to the natural flow of Thought. The former is a natural proceeding, the latter is altogether artificial.

It is not difficult to attract the attention of a child to certain kinds of things, but it is quite impossible to retain it for any length of time. Even when by honest effort, in the way of willing obedience, a child tries his best to attend, and seems to be attending, his teacher may often find that his mind has wandered, and that it has entirely, or at least partially, lost the lesson. Sustained attention is a form of control which even adults have more or less difficulty in mastering. The power of fixing the attention for any length of time effectively on any but the most attractive subjects, is one which few persons possess, and which can only be acquired by practice and considerable self-control. It is, therefore, unreasonable to expect such an accomplishment in children, especially at the age we are now considering. Yet, it is probably one of the commonest mistakes of teachers generally to expect such attention to be promptly given and fixedly maintained by young children.

We may here repeat that attention cannot be *taught*. It is a habit, and must be acquired by experience. Control and concentration are easy if they are habitual, impossible otherwise. Control can only be maintained by constant exercise. Persons who let their Thoughts wander, and who do not habitually fix them on the work of the moment, need not hope to control them when they happen to have that humour. The mind can always be acted upon, whether or not it be under control, and, therefore, the *laissez-faire* habit is destructive of control. Attention cannot be commanded, as it may be in such case distracted either by objects without or suggestions from within. The only possible way of acquiring effective control is by the habitual exercise of the natural power, which is best done by entering into, encouraging, and sympathizing with the early efforts of children, whereby the attention they volunteer is confirmed, and they are thus encouraged to think of and attend to whatsoever they may have in hand. Occupation is of special use in this respect, and helps to keep Thought engaged, instead of allowing it to be captivated or to brood over the suggestions of the lower nature, which are always ready to obtrude themselves on an idle mind.

The art of teaching is in entering into the condition of the pupil's mind, in appreciating his standpoint, and in thoroughly comprehending his ignorance. The success of teachers of dumb animals is largely due to the fact that they expect nothing of their pupils.

It is not easy for a teacher to perceive that things which have always been perfectly familiar to himself, ever since he can remember, can be utterly strange to his pupil. He can hardly help accusing the pupil of either inattention or stupidity. When, however, in addition to the effort necessary to understand his teacher's meaning, the unhappy pupil has, beside the concentration of his mind, to bear the impatience of a teacher who does not see his difficulty, the case becomes distressing, and lessons odious. The ordeal of learning, under such conditions, becomes such as a sensitive child may naturally regard with dread and loathing.

Considering how much of what men learn, not in childhood only but all through life, is acquired by means of the natural method of observation and inference, without any teaching, and when it is further considered how soon that which is acquired by tuition is lost, excepting so much as is kept up by constant use, the rule naturally suggested by these considerations for the first efforts to teach young children is :—

7. To *improve opportunities whensoever they naturally arise*, and *to invite them by attractive means, whereby the mind may be at once informed and trained to attention.*

This being the means by which the greatest proportion of the knowledge men possess is obtained, the most successful for acquiring it and the most effective for enduring effect, it is the means to be most sedulously employed. Lessons are, in fact, attractive or repulsive according as they are well or ill-timed, given when they are wanted, or when they have to be imposed, respecting matters of present interest, or of indifference. The choice of opportunity and the manner in which it is treated, whether judiciously and skilfully, or across the grain, determine the practical use of the lesson, of whatever nature it may be. Supposing that the subject be of present interest, and that attention is thus enlisted, the lesson should nevertheless be *short*. This is very important, because a long one would almost certainly be partly lost, or wholly confused, and also because the principal strain in learning is in the sustained attention it requires. Like food, lessons should be measured to suit the digestive powers. A weak stomach and a young mind have many points of resemblance. As surfeit afflicts the former, so also an overdose of even the most salutary lesson defeats its purpose, and does positive harm to the latter. Surely it is better to have one clear idea, thoroughly and durably impressed on the mind, that is, effectively learnt, than to have two or three confusedly mixed together, like a ravelled thread.

The child in exercising his increasing powers, and especially the free and constant use of his tongue, will afford quite as many opportunities as a tutor could desire. They will be of constant recurrence. Whensoever attention and present interest are manifest they offer an opportunity, whether for correction, suggestion, explanation, or for whatever other form of lesson may be appropriate, to be improved, abused, or lost. To neglect is to lose it. To try to do too much is to abuse it. The way to improve and turn it to the best account is to make it serve for some one brief, opportune, and effective lesson.

In the improvement of opportunities, encouragement is a most effective stimulant. Children love to be noticed and, above all, to be encouraged and applauded. Success is a powerful aid to renewed effort, and should always be noticed. It affords an occasion for creating a joyous emotion, and adding a little sweetness to the cup of life. Moreover, it is the best of all means for making corrections acceptable, and for divesting them of any other than their own proper character. Correction, or aid, should be so adjusted as not to wound reasonable sensibilities, or to seem like needless interference, of which children are generally very sensitive. The golden rule is never more appropriate than in the treatment of children.

CONCLUSION.—The principles we have endeavoured to establish in this work themselves suggest the methods to be adopted for their effective application to the training and management of the infant mind. We have formulated the leading ones into rules, however, in order to show that the system we propose is as simple and reasonable in practice as it is natural in theory. The treatment we recommend is derived from Scripture authority, is in harmony with natural laws, and is confirmed by the results of practical experience.

Beginning at the beginning, we employ the opportunities afforded by the natural conditions proper to each stage of the individual progress, for the purposes for which those conditions are specially suitable, and for which they were evidently designed by Providence. Availing ourselves of them to the fullest extent in each period, we make the best possible preparation for the one next succeeding. Hence, the results of the whole course culminate in the period of Youth, and produce a condition of preparedness the most favourable for the special work of systematic education proper thereto. There our functions terminate. Our business is with the incipient stages of the mental progress. When the mind has attained its maturity it is independent of any such service as we could render. Adults who feel disposed to commence a new discipline after their habits of thought have become fixed, and their mental condition established, would find far better aid than any we could render in existing works on mental philosophy.

If honestly and perseveringly pursued, our methods must necessarily produce in the juvenile mind a harvest corresponding to the seed sown. In the sensitive and susceptible state of personal feeling which dominates in babyhood, we find the special soil in which to sow the seed of good temper, by means of consistent and constant regard to the conditions which are then so exigent. In the pressing wants and appetites of that stage of the infant history, we see a certain means of inspiring his love and trust; and in the state of the babe's helpless dependence his tutors have, *for a time*, the keeping and control of his Will, and may then mould it by unfelt constraint to voluntary submission. We thus introduce the babe to his novel conditions of life under the most favourable auspices, and establish in his mind a happy state of feeling towards the world generally.

During his babyhood, he is thus prepared and equipped for the beginning of a period of more independent activity, during which the seed previously sown will be sedulously cultivated. His agreeable relations with the world may be confirmed by the same means which first produced them, and a docile, amiable temper be developed; or, at least, all that is possible will have been done towards that end.

The third period, or childhood, finds our little subject furnished with a stock of experience of the outward aspects of things acquired by his activities and observation, a free use of spoken language, and also, we hope, with a confirmed good temper and docile disposition, cultivated affections, confidence in tutors, and ideas of moral principles which are good as far as they go. These qualifications prepare the way for the proper use of the newly acquired attainment of speech, and for the new kind of knowledge to be thereby acquired. The natural impulse to learn is now manifested by inquisitive tendencies, which are exactly the means adapted to extend his knowledge beyond the sphere of his own observation. Language enables him to profit by the accumulated experiences of mankind.

With such preparation the youth, when he enters on our fourth period, will possess the qualifications and conditions of mind most favourable for school-work—that is, for receiving, digesting, and assimilating the mental nourishment and discipline of systematic education. If then he have a distinct and desired end in view, a definite object before him to inspire interest and supply adequate inducement for exertion, he will work intelligently, and most probably with willingness and purpose, if not with positive alacrity and cheerfulness.

Our rules will be regarded by some readers as Elisha's too simple prescription was treated by Naaman. They may have expected us to disclose some new discovery in the art of mental discipline, and will therefore be disappointed to find only the old and simple methods rehabilitated. To such persons our answer would be the same which the Syrian's servants addressed to their master. To others our rules may seem to set up a standard of conduct and example which can neither be attained nor hoped for. To this objection we urge that the standard is not ours, but is that of Christian duty, which, whilst admitting that it may not yet be attained, requires that nevertheless it should be the ideal and aim. Our rules do not set up any new standard either of principle or of conduct, nor do they introduce any new difficulties in practice. They do but plead on behalf of infants the same consideration which is already vouchsafed as a matter of course to adults.

We do not suppose that those who are habitually careless of the example they exhibit to adults and elder children would be influenced by our rules to exercise special care regarding their conduct to infants. But there are many tutors who, whilst observing great care in their conduct and example generally, are inconsiderate with regard to infants, in whose presence they act and speak without any such precaution as our rules suggest. To such our arguments may afford convincing proof of their mistake and of its consequences, and thus induce a degree of precaution not hitherto observed.

Others, who have not recognized in the special conditions of the successive stages of the mental and physical progress of infant life the special opportunities afforded for the purposes they serve so well, may be induced to consider those conditions and utilize them in the manner suggested. If those appointed times were more generally understood they would be more effectively observed.

Throughout the foregoing pages the universal and necessary presence of evil in the world is fully admitted and duly taken into account. Where it is not actually expressed it is clearly implied that imperfection, both in character and conduct of tutors, is an inevitable condition of the present order of things. The injunctions contained in our rules cannot be perfectly observed, nor is the command which is addressed to every Christian to "be perfect" practicable by man in his present condition. But though perfection is unattainable it is nevertheless the proper standard towards which human conduct should be ever aimed until that great consummation shall have been attained when God's will shall be done on earth as it is in heaven. This prayer divinely taught, and which every Christian daily utters, imposes on every one the sacred obligation to contribute his part in the great work by which the triumph of good over evil shall be so completed that evil shall be known only to be eternally eschewed by the redeemed race of mankind. Of all the human instrumentalities that can be brought to bear on the furtherance of that great work, none seems to us more promising, more reasonable, or more accordant with Divine injunction, than that of beginning with each individual to direct his Will and faculties aright from the beginning.

APPENDIX.

NOTE 1.—"Blackthorn" was a fine, high-spirited charger, the property of a field-officer in India. He was a good-tempered, valuable horse, and a capital roadster. We knew him both before and after he had been spoiled by bad treatment. When his owner left the country the horse was sold to an old English gentleman, who had a son, a mischievous boy of about fourteen years old. The horse was kept in a loose box, and the boy used to go there and tease him. Keeping himself safe behind the side of the stall, he tormented the animal in this situation, and amused himself by exciting him. After a time the horse became quite furious at the sight of a white face, but continued to be quite tractable with natives. When this intolerance of white people was found to be a settled habit, he was again sold, and the purchaser, a European, trusted to being able to reform him. In this, however, he did not succeed, and in spite of habitual precaution, he had several narrow escapes from the implacable fury of the beast.

On one occasion the animal was sent for the use of the writer of this narrative for a stage of fifteen miles over extremely rough and dangerous paths. His groom, a native, presented a note describing the antipathy the animal had to white people, and further stating that he would do the journey well provided the rider did not dismount and show his face. Of course the precaution was duly observed, the animal did his work splendidly, carrying his rider over rocky streams and the most precipitous paths, where it was most perilous to ride and impossible to dismount. A finer hack could not well be found, and to see him led like a lamb by his native groom, or by any native, he seemed as tractable and good-tempered as ever.

On one occasion when the owner had put him up at an hotel, he became the subject of conversation at table. A gentleman present protested that such ferocity as was attributed to him was impossible. Withdrawing shortly afterwards he went to the stall where the horse was stabled. The horse rushed at him openmouthed and seized him by the middle. Fortunately the horse got hold of him by his cumberband or waist-sash, and was about to paw him with his forefoot when the owner came up and rescued the sufferer.

This case is cited to show the effect of persistent provocation in producing temper, and is valuable as evidence of the fact where no other cause existed, and where the effect was restricted to a certain class of persons identified by the one characteristic they had in common with his tormentor. The case proves incontestably that treatment alone can transform a naturally and habitually good temper into one of the extremest opposite character.

It has already been mentioned that the last owner of "Blackthorn" hoped to conquer him by force and that he was unsuccessful. It remains to be mentioned that, though the cudgel he always carried had the effect of restraining the fury of the beast towards him, it had no influence on the temper of the animal unless to exasperate it, for when his owner was off his guard the horse on several occasions seized the opportunity to attack him. Once, when being led down a precipitous path where both the horse and his owner needed the utmost care to keep their footing, Blackthorn struck out at his master with his forefoot, and the two had a narrow escape of their lives.

This case affords one of many proofs in our experience that coercive measures are both unsafe and ineffectual in the treatment of temper.

NOTE 2.—"Juniper" was one of the finest tempered horses we have ever known, with great power and courage, and no fault we could discover during several years of very close acquaintance. He would go anywhere or do anything within practicable bounds, and was very often severely tried in the long journeys and terrible jungle paths over which he travelled. In proof of his gentle temper it may be mentioned that on one occasion, on his return home after a journey, his rider threw the rein on his neck and left him standing at the foot of the verandah steps whilst he went to greet his wife. Turning short round, however, he found "Juniper" had followed him up the steps and was close behind.

He was sometimes led by his topknot into the dining-room round a table full of guests from whom he would take bread or plantains.

Observing that a very unusual irritability and peevishness began to show itself in him, so different from his well-known habit, we made inquiry for a cause, but at first without success. At length he was developing a really bad temper and showed frequent signs of resentment. He became troublesome to mount, and fidgetty when mounted. At last, observing him one day being groomed, we at once guessed the cause of the change of temper. His groom was using the currycomb, which the horse could not endure. He had a very sensitive skin, peculiarly liable to gall, and he was now being regularly irritated in his most susceptible point. The discontinuance of the offending instrument soon restored his old state of feeling, and we learned a lesson from him, teaching that the best and most settled temper of long standing could be easily destroyed by persistently irritating susceptible feelings.

Judged by this unmistakable sequence of cause and effect, what might be expected of daily scrubbings and dressing tortures, repeated for two or three years on the sensitive natures of young infants, without mention of soap in the eyes and nose, and dirty water in the mouth inadvertently opened in vain remonstrance?

NOTE 3.—" Neptune" was a very large sagacious Newfoundland dog. His master's premises were bounded at one end by a canal, at a considerable distance from the house. Strolling one day on the canal side, Neptune's master fished out some chips which were floating on the water from a neighbouring timber-yard. He amused himself thus for a short time and left the spoil on the bank. A few days later, he received a very unpleasant form of accusation of being privy to a systematic theft. On inquiry he found that "Neptune" had gone into the chips business on his own account. Not content with all he could find floating in the canal he began work at the yard, and when the discovery was made he had made an immense heap of chips, laths, &c., on his master's premises. He had in fact not only imitated, but considerably improved upon, his master's example, and had got his master into trouble.

NOTE 4.—We recently witnessed a very patient attempt to teach a child to tell the time by the clock. Mamma began her explanation in detail of the figures, the hands, the hours and the minutes. She used the simplest terms and gave a complete account of the whole proceeding. The child appeared to be attending, and indeed her face wore an expression of some earnestness during the explanation. The result, however, was a total failure. Mamma had in fact made no visible progress in imparting the knowledge intended, for the child made the wildest answers on being questioned, and showed that she had really no notion whatever of telling the time, but had very confused ideas of all the elements in the process. Mamma appealed to us as witness of the utter stupidity of the child. In reply we gave full credit to the lady for the clear explanations she had given, but we ventured to plead on behalf of the child, that each of her very explicit sentences involved important factors which needed to be understood clearly and remembered, both absolutely and in their relation to the process, before they could be applied. The clock face itself is not quite simple until it is familiar, as its figures, divisions and their significance all involve points of novelty to a child. The hands, too, have not quite simple movements. Finally the explanations were given when the child had no particular interest in them, nor appetite for them. The process is enough for several lessons, and unless some at least of the factors were known before, no child of six could be reasonably expected to learn it all by one effort.

The incident is given as typical of several very common kinds of mistake in teaching—(1) That of giving lessons containing much more than could be learnt at once, whereby confusion of the elements is produced: (2) That of giving lessons when there is no appetite or proper occasion to make them acceptable: (3) That of neglecting to make each point good before proceeding to another.

NOTE 5.—A tutor takes on his knee a little fellow of four or five years old to show him a picture-book. Whilst this is proceeding happily a sweet little cherub of two or three is brought into the room. Seeing his elder brother on the tutor's knee, the little cherub, excited by jealousy, is suddenly transformed into a much lower type of being. Screaming and pulling, he angrily demands his elder brother's dethronement. To prevent a scene the well-behaved brother is deposed and the young despot is taken on the coveted lap. Naturally, after that is done, baby's demands increase, the elder brother must not be allowed even to look on. Baby will have the book as well as the lap!

Of course a single instance of this kind will not complete the education of a spoiled child, but it is thus that those domestic nuisances are produced, and such cases are therefore entirely unjustifiable.

It is needless to say that baby should have been removed and readmitted only on condition of good behaviour. There was a good opportunity lost to him, and a piece of injustice done to the elder brother.

NOTE 6.—In personal intercourse the habit of observing the facial expression of the person addressed is universal, and becomes almost involuntarily the guiding influence of the speaker. An indication of attention and interest on the part of the listener stimulates the speaker, and an opposite expression of indifference or impatience discourages further remark. Thus a vacant look suggests that the mind of the listener is either otherwise absorbed, or is unconscious of what is addressed to him.

In our younger days we paid a visit to an aged friend who was suffering from partial paralysis, and was in an almost dying state. He was a man of very remarkable mental power, of very extensive reading and of the highest order of intelligence. We had enjoyed intimate relations with him before his illness, and now saw him for the first time since. Seeing his countenance, with a strange expression of vacancy, so different from its habitual intelligence and animation, we had no doubt that his reason had fled. Influenced by the expression of his countenance, we spoke to him but little, and that little was dictated by a feeling of polite pity for his mental condition, rather than with any idea that our words were understood. A feeble and unintelligible response indicated a fruitless effort on the part of the invalid, and confirmed our impression of his entire loss of reason.

A few days later, on inquiring of his nurse as to his condition, we were astonished and distressed to learn that he had not only understood our remark, but had also perceived our mistake, and was quite conscious of our having regarded him as being incapable of understanding.

This incident throws some light on the habit of regarding young children as unconscious because of their not having acquired the power and habit of facial expression, and in the absence of which they appear vacant and unconscious.

NOTE 7.—The following incident is given to show that sensations may be perceived and even elicit responsive action without being formulated in the mind, or leaving any durable impression there. It is typical of a great many common experiences which prove that the formulative faculty which fixes perceptions in the mind as durable ideas is a separate act, independent of mere perception.

A merchant whose habit was to keep his private papers in an office box which he took with him in his carriage to and from his office, was asked one evening by his wife, on his return from business, if he had brought a letter for her, as she had requested a friend to send one to the office. "No," was the reply. The wife, surprised and disappointed, repeated her inquiry thinking her husband must be mistaken. "No letter for you has been delivered at my office," was the confident reply. On the following evening the same inquiry was repeated and with the same result. The wife informed her husband that the letter she expected contained money, and that her friend had faithfully promised it should be sent to his office in time for a special purpose. Still he assured her that no such letter had reached him. On the third day the wife sent a special messenger to her friend who lived some miles away to inquire about the missing letter, and was informed that it had been sent according to promise on the day first mentioned, and had been given to her husband himself at his office. In the evening therefore she repeated her inquiry with the same result as before. On the day following, the sender's husband called at the merchant's office and brought the servant who had been entrusted with the letter. Looking intently at the man (a native), the merchant declared he had never seen that man before. "No sir," said the messenger, "you did not look off your book, but you took the letter out of my hand and threw it into your office-box. "Then," said the merchant, "if that be so the letter must be there still." Forthwith he began to turn out the contents of the box, confidently expecting to convict the messenger, but, behold there was the letter! It had slipped down the side and was there a witness against himself. Even as the man narrated the circumstance no trace of recollection came, nor could the merchant recall a single scrap of the incident. But he well remembered the absorbing computations in which he had been engaged that day, and there remained no doubt whatever of the truth of the messenger's statement.

There is nothing very remarkable about the incident, but it is a useful illustration, because the merchant was a methodical man, of excellent memory, and was not at all likely to be mistaken, under ordinary circumstances. Moreover, the whole mystery was, in this case, so completely cleared up, that there is no room to doubt any part of the story, or its moral.

The following illustrate reasoning power in the lower animals, originated by themselves without any teaching :—

NOTE 8.—The sagacity of elephants is well known, and is proved by numerous exhibitions in this country, as well as by the regular services in which they are employed in India and elsewhere. In all these instances, however, the animals have been subject to training and discipline, and therefore they do not show the unaided or natural power of reason. The following illustrate this natural power. At a large kraal, about 1862, in the Kurunegala district of Ceylon, forty-four elephants were captured. Of these a certain number were selected for Government use, and they were accordingly noosed and tied to trees in different parts of the enclosure. Others, to the number of about thirty, were left loose, and were mostly huddled together in a dejected condition. One of the captives was tied to a tree about thirty yards from the group, and as he was a formidable beast, he was tied very firmly and allowed very little tether. Whilst standing near we noticed one of the loose ones out of the herd come stealthily up to the captive to try to unloose him. For this purpose he first tried the strength of the rope, but this was proof. He then felt the knot, and began very vigorously to pick and pull at it. A mahout perceiving this came up and drove him off by spearing him in the heel. Shortly afterwards, the same elephant returned to his friend, and this time made straight for the knot, which he would certainly have unloosed, if he had not been again driven off. As soon as another opportunity offered he again returned, but though he repeated his visits half a dozen times he was never allowed time to complete his work. This elephant and all those taken in the kraal were in their wild state, and had probably never seen a human being before, unless to shun him.

NOTE 9.—Sir J. Emerson Tennent in his work on Ceylon states —as a proof of singular want of sagacity in the wild elephant— that he never attacks the rider when he is encountered by the decoys. In this, however, he is mistaken, for we witnessed an instance of such attack when the decoy elephants ranged up one on either side of the captive mentioned in the last note. In this case the captive watched the approach of the decoys intently, and when they were quite close he cleared the three mahouts off the one on his right with his trunk, and would have served those on the other elephant in the same manner had they not slipped off in time and kept behind the decoy.

NOTE 10.—At the same kraal our attention was much attracted by a magnificent animal, whose sleek fine skin and general appearance gave him the air of a gentleman of his kind. He was tied, but had only one tether, and of a good length, probably six yards or more. When the people threw him a cocoanut he did not fling it away or trample it under foot, as the other captives did, nor did he struggle and make a fool of himself like them, but he behaved sensibly, ate his nuts, and kept quiet. Interested in the brute, we inquired of the Adigar when he would be led away, and were surprised to find that he was to be let loose, as he could not be tamed! When the time came to loose him we were at hand to watch the operation. It was evident from the preparations that the mahouts anticipated difficulty. At length two decoys were drawn up in front of him, at a distance of about thirty yards, and they marched in close order, side by side. One was a tusker and both were large elephants. The captive stood motionless but intent and close to his tree, until the enemy came within the length of his tether. He then drew himself up majestically and made a rush at the tusker, forcing him back on his haunches and completely overwhelmed. His comrade joined in the retreat, having no fancy for such treatment. In vain the mahouts goaded them to renew the attack, and in vain the old Adigar inveighed. The gentleman had effectually demoralized them, and stood again close to his tree ready, if necessary, to make the most of his tether. To renew the attack a third decoy was sent for, and then the three were ranged in close order and marched up. The disdain we fancied we saw depicted in the captive's expression may have been ours, not his; but it seemed to us as though contempt was stamped on every feature as he stood perfectly unresisting and motionless to let them do what they pleased.

Here the whole proceeding was replete with reason. The selection of the tusker was well judged, for his tusks would have otherwise been free for offensive use. The retreat, to give the full length of his tether, and the accurate calculations of time and distance in making his onset, were all proofs of strong power of reason. His tremendous resistance, made when resistance was practicable, and his passive contemptuous submission when resistance was unavailing—indeed, his behaviour from first to last, his resignation to the inevitable, and his sensible view of the whole situation—were simply admirable.

NOTE 11.—The crows in Ceylon are very bold. The Dutch Government, in former times, prohibited their destruction, because of their use as scavengers. Of course they are all wild and quite untaught, but their reasoning powers are very remarkable, and are manifested in many ways.

For instance a puppy dog is gobbling up his meal of rice. Immediately two or three crows come up. One dances up very near to the puppy and threatens an attack on the rice. The puppy then snaps at him, and at that moment a confederate swoops upon the dish and secures a beakful before puppy resumes his meal. Then the game is renewed, each crow in turn getting a part of puppy's food.

NOTE 12.—In a large office in Colombo the partners take their lunch in a private room, the window of which opens on the street. When the partners enter, the window is at once thrown open, and at this signal a number of habitués (crows) range on the sill and shutters. Of the regular attendants of this party one was lame, having lost the foot and lower joint of his leg. He uses his stump freely and does not seem to suffer much by the defect. He, however, receives a decided preference in the distribution of the morsels of meat thrown out. These are mostly so aimed that the others have no chance. One day, however, there were two lame crows, and the gait of one of them was peculiar. However, he came in also for special attention, until it was discovered that he was not lame at all, but had two complete legs, one of which he ingeniously kept half hidden below the sill, and showed the upper half only! The cheat was soon discovered, but not till it had succeeded a good round number of times in deceiving the lunch folk.

NOTE 13.—One evening returning home through the Pettah, or native town, we observed a crow with his head on one side looking very knowing. He was perched on the eaves of a veranda just above the street. Presently he made a swoop across the street and picked a small rice cake, a kind of confection, off a basket in a shop opposite. He returned to his place, pushed the cake under a tile to hide it, and resumed his former place and calculating look. By-and-by he made another swoop, got another cake, and disposed of it likewise. The shop opposite was attended by an old woman, who squatted on one side near a fire which was behind her. She was busy making up cakes, and whenever she finished making one she turned round to put it on the fire to bake. The crow, when she turned round to the fire, made his swoop upon her basket or tray, and had evidently been amusing himself in that fashion for a good while, as he was not able to eat any more and was storing the booty for future provender.

NOTE 14.—In a poultry yard of our acquaintance there were two cock chickens of a very large breed that had been sent by a friend, and not bred in the yard. These birds kept apart from the rest and never mixed with them, but strayed away by themselves, and were almost always fighting. They were, in fact, called the Cheeryble Brothers, by ironical allusion, for that reason. There were also in the yard some ducks, one of which was called Pacifex, because she always rushed in between the Cheeryble brothers whenever she saw them fighting, and she drove them apart, generally ejecting one of them out of the gate. Hers was a troubled life, because of the frequent need for her interference, and also because the pugnacious cocks always kept at a distance. One day she was observed waddling with a more than usual rapidity evidently on urgent duty, and she was not this time making for the Cheerybles, but was bent on arresting Sambo, a tame crow, which was making free with the food of an old white hen that was always fed separately, because she was scared by the other fowls and could not contend with them for her position. Beside she was an old favourite and was specially favoured. Now, on the occasion in question, Sambo taking advantage of

Sylvia's timidity, was gobbling up her food, when Pacifex came from the further end of the yard to Sylvia's rescue. Sambo prudently retired as soon as he saw Pacifex approach, and Sylvia took the rest of her food under the protection of the police.

We have since heard that ducks frequently behave in the manner of our friend Pacifex, but she was the only specimen of the kind in our acquaintance.

NOTE 15.—Whilst a lady of our acquaintance was walking in her garden, a large dog entered by the door, which had been left open. Having a great fear of dogs she was not a little alarmed by the apparition of so formidable a stranger, especially as he came directly towards her. He was so extremely lame, however, and seemed to be suffering so severely, that she made no attempt to escape from him. He therefore came up to her, and held up his lame paw with such piteous and beseeching gestures that her sympathy overcame her fears. So she went into the house, and the animal followed her. The wound was deep and angry, and, being also full of dirt, it needed a painful dressing, which the poor dog endured patiently and without resistance. With looks of gratitude he took his leave and went away.

A few days afterwards, the old patient, finding the garden door open, returned, bringing with him a lame companion, which he introduced with unmistakable solicitation for the kind attention he had himself formerly received. He was now quite well and walked with his natural gait. Our lady friend attended to the new patient, and in dismissing him and his cicerone gave explicit orders that the garden door should in future be kept shut, lest her house should become a regular hospital for canine invalids.

NOTE 16.—Passing along my veranda one morning whilst living in Ceylon, I observed a train of large black ants marching at a business-like pace into a hole in the floor. A train of ants is such a very common occurrence that the one in question excited but little notice. Having occasion to pass the same place four hours later I found the train was still on the march, and my servant informed me that it had been going on uninterruptedly since morning. The train was about two inches wide and the ants may have been six or eight abreast in close order. There was not a gap in the ranks. The train extended across my premises into a jungle, whither I could not further proceed. The train all this length was unbroken, the order perfect, and the march was continuous for six hours at the least.

The hole into which the whole army entered led into an air passage, and thence under my drawing-room floor, where there was a space of about two feet deep. As the air passage was but a few inches wide, it was not possible to discover what was going on under the floor or what attracted this tremendous host, so the incident passed out of mind, and would not have been again thought of if it had not been for what followed.

In the evening after dinner, whilst we were playing a concerted piece, my wife started with a shudder and pointed to a large cockroach on the wall. His staggering gait and struggling movements revealed the fact that he was beset by ten or a dozen ants. Stopping our music we discovered that there were hundreds of cockroaches on the walls of the room all similarly beset with ants. Some took to the wing, but their flight was quite as spasmodic as their other movements. It was then evident that a colony of cockroaches must have taken up their quarters under the floor, and that the ants had discovered their fortress, and besieged and taken the whole garrison prisoners.

The whole of this proceeding appears to us replete with evidence of *reason*. It seems to be as much a matter of reason as Lord Wolseley's march to Tel-el-Kebir and his notable victory there. The train of ants maintained its order and its motion unbroken, although it crossed a road and must have suffered considerable loss by the passage of my carriage and horses several times during the march. There must have been a general and a plan. The skill of the conception must have been admirable to enable wingless ants to besiege and capture such a host of powerful insects with strong power of flight. As the siege was conducted in the dark no idea of the plan could have been ascertained, but it was necessarily concerted, required every member of the invading host to take an appointed part, and was perfectly successful, for we have reason to believe that the capture was complete. Even those who believe in instinct must see that this proceeding exceeds the widest bounds that could be assigned to such a faculty. There were plan, combination, and *communication* all plainly and undeniably involved in the work and its execution, and the whole campaign was conceived and carried out by creatures that had no help from human teacher or example.

APPENDIX. 413

NOTE 17.—List of words alluded to in page 147.

A	Bird	Cart	Cream	Early
About	Bite	Carry	Crooked	Ear
Above	Black	Cat	Cross	Easter
Ache	Bless	Caught	Crow	Easy
Afraid	Blind	Cellar	Cruel	Eat
After	Blood	Chair	Crumb	Edge
Afternoon	Blot	Change	Crust	Egg
Again	Blue	Charge	Cry	Eight
Against	Board	Chatter	Cuckoo	Either
Ago	Boat	Cheat	Cup	Elbow
Ah	Body	Child	Cushion	Elephant
Alike	Bell	Chill	Cut	Else
Alive	Bone	Chimney		Empty
All	Book	Chin	Daddy	End
Alone	Boot	Choice	Dance	Enough
Alphabet	Born	Choke	Dark	Equal
Already	Both	Choose	Darling	Evening
Always	Bottle	Christmas	Day	Ever
Amen	Bottom	Church	Dead	Every
Almost	Bow	Claw	Deaf	Except
And	Branch	Clean	Deal	Excuse
Angry	Bread	Climb	Dear	Expect
Answer	Break	Cling	Declare	Eye
Apple	Breast	Cloak	Deep	
As	Breath	Clock	Delight	Face
Ask	Bridge	Close	Diddle	Fair
Asleep	Bright	Closet	Did	Fairy
Ass	Bring	Clothes	Die	Fall
At	Broke	Clumsy	Dig	Falsehood
Aunt	Brow	Coal	Dinner	Family
Awake	Brown	Coat	Ding-dong	Fan
Away	Brush	Coax	Directly	Far
	Bud	Cock	Dirt	Farther
Baby	Bull	Coffee	Dish	Father
Back	Bump	Cold	Dizzy	Fast
Bad	Bun	Collar	Do	Fasten
Bag	Boy	Colour	Doctor	Fat
Ball	Burn	Comb	Dog	Fault
Bang	Busy	Come	Doll	Fear
Bare	But	Comfortable	Done	Feather
Bark	Butcher	Common	Donkey	Feed
Basin	Butter	Confess	Door	Feel
Basket	Button	Contrary	Dot	Feet
Bath	Buy	Coo	Dove	Fell
Bead		Cook	Down	Fender
Bear	Cake	Copy	Draw	Few
Best	Calf	Copybook	Dreadful	Fiddle
Beauty	Call	Cord	Dress	Fidget
Because	Came	Corner	Drink	Fir
Begin	Can	Cot	Drop	Field
Beside	Canary	Cotton	Drum	Fifteen
Best	Candle	Cough	Dry	Fig
Better	Cap	Count	Duck	Fight
Between	Card	Cousin	Dunce	Fill
Bit	Care	Cover	Dust	Find
Bible	Carpet	Cow		Fine
Big	Carriage	Crack	Each	Finger

Fire	Go	Hiss	Kind	Lump
First	Goat	Hit	King	
Fish	Gobble	Hold	Kiss	Made
Fist	God	Hole	Kite	Make
Fit	Gone	Home	Knee	Man
Five	Good	Hook	Kneel	Mamma
Fix	Goose	Hoop	Knit	Manner
Flannel	Got	Hop	Knob	Many
Flat	Grace	Hope	Knock	Mark
Flesh	Grass	Horrid	Knot	Mat
Floor	Grate	Horse	Know	Match
Flour	Gravy	Hot		Matter
Flower	Grease	Hour	Lady	May
Fly	Great	House	Laid	Mean
Fold	Greedy	How	Lame	Measure
Follow	Green	Hump	Lamp	Meddle
Food	Grope	Hungry	Landing	Medicine
Foolish	Ground	Hurrah	Large	Meat
Foot	Grow	Hurry	Lark	Meet
For	Growl	Hurt	Last	Memory
Forget	Grunt	Hush	Lake	Mend
Forgive	Guess		Laugh	Mess
Fork	Gun	I	Lay	Mew
Form		Ice	Lead	Middle
Forward	Hail	Idle	Leaf	Middling
Foul	Hair	If	Learn	Might
Found	Hairy	Ill	Least	Milk
Four	Half	Impossible	Leave	Mince
Fox	Hammer	Impudent	Left	Mind
Free	Hand	In	Leg	Mine
Freeze	Handkerchief	Indeed	Length	Minute
Fresh	Handle	Ink	Less	Mischief
Fretful	Hang	Insect	Let	Miss
Friday	Happen	Inside	Letter	Mistake
Friend	Happy	Instead	Lid	Mite
Fright	Hard	Into	Lie	Mix
Frill	Hardly	Iron	Life	Monday
Frock	Hark	Is	Lift	Money
Frog	Harm	It	Like	Moon
From	Hat	Itch	Likely	More
Front	Have	Itself	Lion	Morning
Frost	He		Lip	Most
Fruit	Head	Jack	Lisp	Mother
Full	Heap	Jam	Listen	Mouse
Fun	Hear	Jar	Litter	Mouth
Fye	Heat	Jelly	Little	Move
	Heaven	Job	Live	Much
Game	Heavy	Joint	Loaf	Mud
Garden	Height	Jolly	Lock	Music
Gas	Held	Joy	Loll	Must
Gate	Help	Jug	Long	Mutton
Gather	Hem	Jump	Longing	My
Gentleman	Hen	Just	Look	
Giant	Her		Loose	Naked
Giddy	Here	Keep	Lost	Name
Gift	Herself	Kettle	Lot	Narrow
Give	Hide	Key	Loud	Nasty
Glad	Him	Kick	Love	Naughty
Glass	His	Kill	Low	Nay

Near	Past	Pussy	Same	Silk
Neat	Pat	Put	Sand	Silly
Neck	Paw	Puzzle	Sash	Silver
Needle	Pay		Saturday	Since
Neither	Pea	Quack	Save	Single
Nest	Peace	Quarrel	Saw	Sing
Never	Pear	Quarter	Say	Sir
New	Peel	Queen	Scald	Sister
Next	Peep	Question	Scatter	Sit
Nice	Peg	Quick	School	Six
Night	Pen	Quiet	Scissors	Size
Nine	Pencil	Quite	Scold	Skin
No	Penny		Scrap	Skip
Nobody	People	Rabbit	Scrape	Skirt
Noise	Perhaps	Race	Scratch	Sky
None	Pet	Rag	Scream	Slap
Nonsense	Petticoat	Rail	Scrub	Sleeve
Noon	Pick	Rain	Second	Sleep
Nor	Pie	Rap	Secret	Slice
Nose	Piece	Rat	See	Slide
Not	Pig	Raw	Seed	Slip
Nothing	Pigeon	Reach	Seek	Slipper
Now	Pillow	Read	Seem	Sly
Number	Pin	Ready	Seen	Smack
Nurse	Pinch	Red	Self	Small
Nursery	Pity	Remember	Selfish	Smell
Nut	Place	Repeat	Send	Smile
	Plain	Rest	Sense	Smoke
Of	Plant	Rib	Servant	Smooth
Off	Plate	Ribbon	Set	Snail
Often	Play	Rice	Seven	Snake
Oil	Pleasant	Riddle	Sew	Snatch
Old	Please	Ride	Shake	Sneeze
On	Plenty	Right	Shall	Snow
Once	Plum	Ring	Shame	So
One	Pocket	Ripe	Shape	Soap
Only	Point	Rise	Sharp	Soft
Open	Porridge	Road	Shawl	Sock
Opposite	Positive	Rob	She	Sod
Or	Postman	Rod	Sheep	Soil
Orange	Posy	Roll	Sheet	Soldier
Order	Pot	Romp	Shelf	Some
Other	Pour	Roof	Shell	Son
Our	Powder	Room	Shin	Song
Out	Prayer	Rose	Shine	Soon
Over	Present	Rot	Shirt	Soot
Own	Press	Rough	Shiver	Sop
	Pretend	Round	Shock	Sound
Pain	Pretty	Row	Shoe	Soup
Pair	Prick	Rub	Shoot	Sour
Pan	Prize	Rude	Shop	Sow
Parcel	Proud	Run	Shoulder	Spade
Pardon	Provoke		Show	Spare
Parrot	Pudding	Sabbath	Shut	Speak
Part	Puddle	Sad	Shutter	Spill
Particular	Pull	Safe	Shy	Spin
Pass	Puppy	Said	Sick	Spoil
Passage	Purpose	Sake	Side	Spoke
Passion	Push	Salt	Sight	Spoon

Spot	Sweet	Thump	Twinkle	Weigh
Spread	Swing	Thursday	Twist	Well
Spring	Switch	Tickle	Two	Went
Square		Tidy		Were
Squeak	Table	Tie	Ugly	Wet
Squeeze	Tail	Tight	Uncle	What
Squint	Tailor	Till	Under	Wheel
Stairs	Take	Time	Understand	When
Stammer	Tale	Tin	Undo	Where
Stamp	Talk	Tiny	Undress	Which
Stand	Tall	Tired	Unfair	While
Star	Tape	To	Unfasten	Whip
Steal	Tart	Toast	Unkind	Whisper
Step	Tea	To-day	Until	Whistle
Stick	Teach	Toe	Unto	White
Stiff	Tear	Told	Untruth	Whose
Still	Teeth	Tongs	Unwell	Why
Sting	Tell	Tongue	Up	Will
Stir	Temper	Too	Upset	Wind
Stitch	Ten	Took	Us	Window
Stomach	Than	Top	Use	Wing
Stone	Thank	Topsy-turvy	Useful	Winter
Stool	That	Toss		Wipe
Stop	The	Touch	Veil	Wish
Story	Their	Towards	Verse	With
Straight	Them	Towel	Very	Without
Stretch	Then	Town	Vex	Wonder
Strike	There	Toy	Visit	Wool
String	These	Train	Voice	Word
Stripe	They	Tray		Worm
Stroke	Thick	Tread	Waist	Worse
Strong	Thief	Treat	Wake	Worsted
Stuff	Thimble	Tree	Walk	Worth
Such	Thin	Tremble	Wall	Would
Suck	Think	Trick	Want	Wrap
Sugar	Third	Trot	Warm	Wring
Suit	Thirst	Trouble	Was	Write
Sulk	This	True	Wash	Wrong
Summer	Those	Try	Waste	
Sun	Though	Tub	Watch	Yard
Sunday	Thought	Tuck	Water	Yellow
Supper	Thread	Tumble	Wax	Yesterday
Suppose	Three	Tune	Way	Yes
Sure	Through	Turn	Wear	You
Swallow	Throw	Twelve	Weather	Yours
Sweep	Thumb	Twice	Wee	

www.ingramcontent.com/pod-product-compliance
Lightning Source LLC
Chambersburg PA
CBHW020540300426
44111CB00008B/738